ID0996874

THE BIG QUESTIONS
Evolution

Francisco J. Ayala is University Professor and Donald Bren Professor of Biological Sciences and Professor of Philosophy at the University of California, Irvine. He is a member of the National Academy of Sciences (NAS), a recipient of the 2001 National Medal of Science, and served as Chair of the Authoring Committee of *Science, Evolution, and Creationism*, jointly published in 2008 by the NAS and the Institute of Medicine.

He has written numerous books and articles about the intersection of science and religion, including *Darwin's Gift to Science and Religion* and *Am I a Monkey?*. Dr Ayala has received numerous honorary degrees and awards, including the 2010 Templeton Prize for his exceptional contribution to affirming life's spiritual dimension.

The Big Questions confronts the fundamental problems of science and philosophy that have perplexed enquiring minds throughout history, and provides and explains the answers of our greatest thinkers. This ambitious series is a unique, accessible and concise distillation of humanity's best ideas.

Series editor **Simon Blackburn** is Professor of Philosophy at the University of Cambridge, Research Professor of Philosophy at the University of North Carolina and one of the most distinguished philosophers of our day.

Titles in *The Big Questions* series include:

PHILOSOPHY
PHYSICS
THE UNIVERSE
MATHEMATICS
GOD
EVOLUTION

THE BIG QUESTIONS

Evolution

Francisco J. Ayala

SERIES EDITOR
Simon Blackburn

Quercus

Contents

INTRODUCTION

The eminent evolutionist Theodosius Dobzhansky famously asserted in 1973 that 'Nothing in biology makes sense except in the light of evolution.' Indeed, the theory of biological evolution is the central organizing principle of modern biology. Evolution provides a scientific explanation for why there are so many different kinds of organisms on Earth, and gives an account of the similarities and differences in their appearance, as well as in their genetic make-up and physiology. It accounts for the origin of humans on Earth and reveals our species' biological connections with other living things. It provides an understanding of the constantly evolving bacteria, viruses and other pathogenic organisms, and enables the development of effective new ways to protect ourselves against the diseases they cause.

Knowledge of evolution has made possible improvements in agriculture and medicine, and has been applied in many fields outside biology; for example, in software engineering, where genetic algorithms seek to mimic evolutionary processes, and chemistry, where the principles of natural selection are used for developing new molecules with specific functions.

Charles Darwin is deservedly recognized as the founder of the theory of evolution. In *On the Origin of Species*, published in 1859, he laid out the evidence demonstrating the evolution of organisms. More important yet is that he discovered natural selection, the process that accounts for the 'design' of organisms and for their diversity. Darwin's *Origin of Species* is, first and foremost, a sustained effort to solve the problem of how to account scientifically for the adaptations of organisms. Darwin seeks to explain the design of organisms – their complexity, diversity and marvellous contrivances – as the result of natural processes.

Darwin and other nineteenth-century biologists found compelling evidence for biological evolution in the comparative study of living organisms, in their geographic distribution and in the fossil remains of extinct organisms. Since Darwin's time, the evidence from these sources has become stronger and more comprehensive, while biological disciplines that have emerged recently – genetics, biochemistry, ecology, animal behaviour (ethology), neurobiology and especially molecular biology – have supplied powerful additional evidence and detailed confirmation. Accordingly, evolutionists are no longer concerned with obtaining evidence to support the fact of evolution. Rather, evolutionary research nowadays seeks to understand further, and in more detail, how the process of evolution occurs.

The theory of evolution is perceived by many people as controversial. I find this perception surprising. It is beyond reasonable doubt that organisms, including humans, have evolved from ancestors that were very different from them. The evolution of organisms is accepted by scientists with the same degree of confidence as they accept other well-confirmed scientific theories, such as the revolution of the Earth around the Sun, the expansion of galaxies, atomic theory or the genetic theory of biological inheritance.

Molecular biology, a scientific discipline that came about in the second half of the twentieth century, a hundred years after the publication of *Origin*, has made it possible to reconstruct the 'universal tree of life', the continuity of succession from the original forms of life, ancestral to all living organisms, to every species now living on Earth. The virtually unlimited information that is encoded in the DNA sequence of living organisms allows evolutionists to reconstruct in detail all evolutionary relationships leading to present-day organisms.

The Big Questions: Evolution advances thoughtful reflections about 20 questions that are often raised by those who want to know about evolution. These questions are among the most important to the theory of evolution, though many others have, of necessity, been left out. The treatment is accurate, and profound when appropriate, although it is not intended for the expert, but rather for general readers who may have no specialist knowledge but who are assumed to be interested and intelligent.

WHAT IS EVOLUTION?

The central-unifying concept of biology

*T*he evolution of organisms is at the core of biological disciplines such as genetics, molecular biology, biochemistry, neurobiology, physiology and ecology, and makes sense of the emergence of new diseases, the development of antibiotic resistance in bacteria and other matters of public health. Evolution explains the agricultural relationships among wild and domestic plants and animals. It is used in informatics and computer science, in the design of chemical compounds and in various other industries.

The term 'evolution' means change over time. It usually refers to the evolution of living things, but it is also used in other scientific contexts, particularly in astronomy to refer to the processes by which the whole physical universe – galaxies, stars and planets – forms and changes.

Contrary to popular opinion, neither the term nor the idea of biological evolution began with Charles Darwin and his foremost work, *On the Origin of Species by Means of Natural Selection* (1859). The *Oxford English Dictionary* (1933) tells us that the word 'evolution', to unfold or open out, derives from the Latin *evolvere*, which applied to the 'unrolling of a book'. It first appeared in the English language in 1647 in a non-biological connection, and became widely used in English for all sorts of progressions from simpler beginnings. Evolution was first used as a biological term in 1670, to describe the changes observed in the maturation of insects. However, it was not until the 1873 edition of *Origin of Species* that Darwin first employed the term to refer to what we now call biological evolution. He had earlier used the expression 'descent with modification', which is still a good brief definition of the process.

The universe

It is widely believed that the universe started about 15 billion years ago in the Big Bang, a monumental explosion that sent matter and energy expanding in all directions. As the universe expanded, matter collected into clouds that gradually condensed into galaxies such as our own Milky Way. In these galaxies gravitational attraction compressed the material, which in many cases condensed into stars, where nuclear reactions took place. In the case of our Sun, gas and dust collided and aggregated, forming very small planets, which in successive stages coalesced into the eight (or nine, if Pluto is included) planets of our Solar System and their numerous satellites.

The age of the Earth is estimated at 4.54 billion years. The oldest known rocks, dated at 3.96 billion years, are found in north-western Canada, although rocks found in other places, such as Western Australia, encase zircon crystals dated at 4.3 billion years, older than the rocks themselves. The origins of life on Earth may have been as early as 4 billion years ago. There is evidence that organisms similar to today's bacteria lived 3.5 billion years ago. All species living today, estimated to number at least 10 million, derive by evolution from those early simple organisms.

Biological evolution

Evolution is the process of 'descent with modification', as Charles Darwin named it, through which all species now living, and the even more numerous ones that became extinct in the past, came about.

The virtually infinite variations on life are the fruit of evolution. All living creatures are related by descent from common ancestors. Humans and other mammals descend from shrew-like creatures that lived more than 150 million years ago; mammals, birds, reptiles, amphibians and fishes share as ancestors aquatic worms that lived 600 million years ago; and all plants and animals derive from bacteria-like microorganisms that originated more than 3 billion years ago. Lineages of organisms change through generations; diversity arises because the lineages that descend from common ancestors diverge through time as they adapt to different environments.

Evolution is the process of 'descent with modification' through which all species now living, and the even more numerous ones that became extinct in the past, came about.

Evolutionary research consists mainly of three different, though related, issues: (1) the fact of evolution; that is, that organisms are related by common descent with modification; (2) evolutionary history; that is, the details of when lineages split from one another and of the changes that occurred in each lineage; and (3) the mechanisms or processes by which evolutionary change occurs.

The fact of evolution is the most fundamental issue and the one established with utmost certainty. Darwin gathered much evidence in its support, but the evidence has accumulated continuously ever since, derived from all biological disciplines. The evolutionary origin of organisms is today a scientific conclusion established with the kind of certainty attributable to such concepts as the roundness of the Earth, the motions of the planets and the molecular composition of matter. This degree of certainty beyond reasonable doubt is what is implied when biologists say that evolution is a 'fact'; the evolutionary origin of organisms is accepted by virtually every biologist.

The second and third issues go far beyond the general affirmation that organisms evolve. The theory of evolution seeks to ascertain the relationships between particular organisms and the events of evolutionary history, as well as to explain how and why evolution takes place. These are matters of active scientific investigation. Many conclusions are well established; for example, that chimpanzees and gorillas are more closely related to humans than are any of those three species to baboons or other monkeys; or that natural selection explains the adaptive configuration of such features as the human eye and the wings of birds. Some other matters are less certain, others are conjectural and still others – such as precisely when life originated on Earth and the characteristics of the first living things – remain largely unresolved.

However, uncertainty about these issues does not cast doubt on the fact of evolution. Similarly, we do not know all the details about the configuration of the universe and the origin of the galaxies, but this is not a reason to doubt that the galaxies exist or to throw out all we have learned about their characteristics. Biological evolution is one of the most active fields of scientific research at present, and significant discoveries continually accumulate, supported in great part by advances in other disciplines.

The study of biological evolution has transformed our understanding of life on this planet. Evolution provides a scientific explanation for why there are so many different kinds of organisms on Earth and how they are all part of an evolutionary lineage. It demonstrates why some organisms that look quite different are in fact closely related, while others that may look similar are distantly related. It accounts for the appearance of humans on Earth and reveals our species' biological connections with other living things. It details how different groups of humans are related to each other and how we acquired many of our features. It enables the development of effective new ways to protect ourselves against constantly evolving bacteria and viruses.

Evolution provides a scientific explanation for why there are so many different kinds of organisms on Earth and how they are all part of an evolutionary lineage.

Brief history

Explanations for the origin of the world, man and other creatures are found in all human cultures. Traditional Judaism, Christianity and Islam explain living beings and their adaptations to their environments – wings, gills, hands, flowers – as the handiwork of an omniscient God. The philosophers of ancient Greece had their own creation myths. Anaximander proposed that animals could be transformed from one kind into another, and Empedocles speculated that they could be made up of various combinations of pre-existing parts. Closer to modern evolutionary ideas were the proposals of early Christian church authors like Gregory of Nazianzus and Augustine, who maintained that not all species of plants and animals were created as such by God; rather some had developed in historical times from other of God's creations. Their motivation was not biological but religious. Some species must have come into existence only after the Flood, because it would have been impossible to hold representatives of all species in a single vessel such as Noah's Ark.

Christian theologians of the Middle Ages did not directly explore the notion that organisms might change by natural processes, but the matter was, usually incidentally, considered as a possibility by many, including Albertus Magnus and his student Thomas Aquinas.

In the eighteenth century, Pierre-Louis Moreau de Maupertuis proposed the spontaneous generation and extinction of organisms as part of his theory of origins, but he advanced no theory about the possible transformation of one species into another through knowable natural causes. One of the greatest naturalists of the time, Georges-Louis Leclerc, Comte de Buffon, explicitly considered – and rejected – the possible descent of several distinct kinds of organisms from a common ancestor. However, he made the claim that organisms arise from organic molecules by spontaneous generation, so that there could be as many kinds of animals and plants as there are viable combinations of molecules.

The Swedish botanist Carolus Linnaeus devised the hierarchical system of plant and animal classification that is still in use in a modernized form. Although he insisted on the fixity of species, his classification system eventually contributed much to the acceptance of the concept of common descent. Erasmus Darwin, grandfather of Charles, offered some evolutionary speculations in his *Zoonomia; or, The Laws of Organic Life*, but they were not systematically developed and had no real influence on subsequent theories.

The great French naturalist Jean-Baptiste Lamarck held the view that living organisms represent a progression, with humans as the highest form. In his *Philosophical Zoology*, published in 1809, the year in which Charles Darwin was born, he proposed the first broad theory of biological evolution. Organisms evolve through aeons of time from lower to higher forms, a process still going on and always culminating in humans. As they become adapted to their environments through their habits, modifications occur. Use of an organ or structure reinforces it; disuse leads to obliteration. We see, for example, that an athlete develops robust muscles, while unused organs, such as the vermiform appendix, gradually die out. The characteristics acquired by use and disuse, according to Lamarck's evolutionary theory, would be inherited. This assumption, later called the inheritance of acquired characteristics, was thoroughly disproved in the twentieth century. Other Lamarckian theories, such as the notion that the same organisms repeatedly evolve in a fixed sequence of transitions, so that, for example, today's primates will in the future have humans as their descendants, have also been disproved.

Charles Darwin

The founder of the modern theory of evolution is Charles Darwin. On 27 December 1831, a few months after his graduation from the University of Cambridge, he sailed as a naturalist aboard the HMS *Beagle* on a round-the-world trip that would last until October 1836. He was often able to disembark for extended trips ashore to explore the local fauna and flora and to collect natural specimens.

> *It was Darwin's greatest accomplishment to show that the complex organization and functionality of living beings can be explained as the result of a natural process — natural selection — without any need to resort to a Creator or other external agent.*

In Argentina he studied fossil bones from large extinct mammals. In the Galápagos Islands he observed numerous species of finches, as well as the giant tortoises after which the islands were named. These are among the events credited with stimulating his interest in how different species arise in different places and different times, and why some species become extinct. In 1859 he published *On the Origin of Species*, a treatise providing extensive evidence for the evolution of organisms and proposing natural selection as the key process determining its course. He published many other books as well, notably *The Descent of Man, and Selection in Relation to Sex* (1871), which provides an evolutionary account of human origins.

Two revolutions: Copernicus and Darwin

The so-called Copernican Revolution was launched with the publication in 1543, the year of Nicolaus Copernicus' death, of his *De revolutionibus orbium celestium* (*On the Revolutions of the Celestial Spheres*), and bloomed with the publication in 1687 of Isaac Newton's *Philosophiae naturalis principia mathematica* (*The Mathematical Principles of Natural Philosophy*). The discoveries by Copernicus, Kepler, Galileo, Newton and others in the sixteenth and seventeenth centuries had gradually ushered in a conception of the universe as matter in motion governed by natural laws. It was shown that the Earth is not the centre of the universe but a small planet rotating around an average star; that the universe is immense in space and time; and that the motions of the planets around the Sun can be explained by the same simple laws that account for the motion of physical objects on our planet.

These and other discoveries greatly expanded human knowledge. The conceptual revolution they brought about was more fundamental yet: a commitment to the postulate that the universe obeys immanent laws that account for natural phenomena. The workings of the universe were brought into the realm of science: explanation through natural laws. All physical phenomena could be accounted for as long as the causes were adequately known.

The advances of physical science brought about by the Copernican Revolution had driven mankind's conception of the universe to a split-personality state of affairs, which persisted well into the mid nineteenth century. Scientific explanations, derived from natural laws, dominated the world of non-living matter, on Earth as well as in the heavens. Supernatural explanations that depended on the unfathomable deeds of the Creator accounted for the origin and configuration of living creatures – the most diversified, complex and interesting realities of the world.

It was Darwin's genius to resolve this conceptual schizophrenia. He completed the Copernican Revolution by drawing out for biology the notion of nature as a lawful system of matter in motion that human reason can explain without recourse to supernatural agencies. The conundrum faced by Darwin can hardly be overestimated. The strength of the argument from design to demonstrate the role of the Creator had been forcefully set forth by theological and other religious writers. Wherever there is design, we look for its author. It was Darwin's greatest accomplishment to show that the complex organization and functionality of living beings can be explained as the result of a natural process – natural selection – without any need to resort to a Creator or other external agent. The origin and adaptation of organisms in their profusion and wondrous variations was thus brought into the realm of science.

Darwin accepted that organisms are 'designed' for certain purposes; that is, they are functionally organized. Organisms are adapted to certain ways of life and their parts are designed to perform certain functions. Birds have wings for flying, fish have gills to breathe in water and trees have leaves to capture the sunlight. But he went on to provide a natural explanation of the design. The design features of living beings could now be explained, like the phenomena of the inanimate world, by the methods of science, as outcomes of natural laws manifested in natural processes that could be subject to test by observation and experiment.

Alfred Russel Wallace is famously given credit for discovering, independently of Darwin, natural selection as the process accounting for the evolution of species. Although his independent discovery of natural selection is remarkable, he was not interested in explaining design, but rather in accounting for the evolution of species, as indicated in the title of his 1858 paper: 'On the Tendency of Varieties to Depart Indefinitely from the Original Type'. Wallace thought that evolution proceeds indefinitely and is progressive. Darwin, on the other hand, did not accept that evolution would necessarily represent progress or advancement, nor did he believe that it would always result in morphological change over time; rather, he knew of the existence of 'living fossils', organisms that had remained unchanged for millions of years. For example, 'some of the most ancient Silurian animals, such as the Nautilus, Lingula, etc., do not differ much from living species'.

In 1858, Darwin was at work on a multi-volume treatise, intended to be titled *On Natural Selection*. Wallace's paper stimulated him to write *On the Origin of Species*, which would be published the following year. Darwin's focus, in *Origin* as elsewhere, was the explanation of design, with evolution playing the subsidiary role of supporting evidence.

WAS DARWIN RIGHT?
Evolution is a fact

'*Scientists use the term 'fact' to refer to a scientific explanation that has been tested and confirmed so many times that there is no longer a compelling reason to keep testing it or looking for additional examples . . . Because the evidence supporting [evolution] is so strong, scientists no longer question whether biological evolution has occurred and is continuing to occur. Indeed, they investigate the mechanisms of evolution, how rapidly evolution can take place, and related questions.*' [1]

Many scientific explanations have been so thoroughly tested that they are very unlikely to change in substantial ways as new observations are made or new experiments are analysed. These explanations are accepted by scientists as being true and factual descriptions of the natural world. The atomic structure of matter, the genetic basis of heredity, the circulation of blood, gravitation and planetary motion and the process of biological evolution by natural selection are just a few examples of a very large number of scientific explanations that have been overwhelmingly substantiated.

Darwin's evolution
'Was Darwin right?' may refer to the fact of evolution; that is, Darwin's evidence in support of evolution and whether evolution has been definitely confirmed by science. The question may also refer to natural selection, the process discovered by Darwin that

explains how evolution occurs and why organisms and their organs and limbs are 'designed' to accomplish certain ways of life. In *On the Origin of Species*, Darwin dedicated Chapters 9–13 to the first aspect, the evidence for evolution, and Chapters 1–8 and 14 to the second, natural selection as the process that accounts for design and adaptation. Science has confirmed that he was right with respect to both questions.

What is Natural Selection? is the subject of the third chapter in this book, and the topic is further explored in *What is Survival of the Fittest?*, and elsewhere. The evidence for evolution considered by Darwin includes palaeontology (the study of fossils), biogeography (the geographic distribution of plants and animals) and comparative anatomy (form) and embryology (how form comes about). We'll review the palaeontology evidence in *What Does the Fossil Record Tell Us?*, and *What is the Missing Link?*. In the current chapter we'll consider biogeography and comparative anatomy and embryology. Later on, *What is Molecular Evolution?*, will bring in the evidence from molecular biology, a scientific discipline and type of evidence that did not actually exist in Darwin's time; it emerged only in the 1950s, following the discovery of the structure of DNA (deoxyribonucleic acid) as the molecule encompassing biological heredity. Molecular biology provides the most convincing evidence for evolution and makes it possible to reconstruct the evolutionary history of all living organisms.

> *Because of the immense advances in the study of evolution, it is now possible to assert that gaps of knowledge in the evolutionary history of living organisms no longer exist.*

Because of the immense advances in the study of evolution, it is now possible to assert that gaps of knowledge in the evolutionary history of living organisms no longer exist. Using molecular biology, scientists have reconstructed the 'universal tree of life' (see *What is the Tree of Life?*), the continuity of succession from the original forms of life, ancestral to all organisms, to every species now living on earth. The virtually unlimited evolutionary information encoded in the DNA sequence of living organisms allows evolutionists to replicate in detail all evolutionary relationships leading to present-day organisms. Invest the necessary resources (time and laboratory expenses) and you can have the answer to any query, with as much precision as you want.

Similarity of form

Humans, horses, mice, whales, bats, birds and turtles have skeletons that are strikingly similar, in spite of the different ways of life of these animals and the diversity of their environments. The correspondence, bone by bone, can easily be seen in the limbs as well as in other parts of the body. From a purely practical point of view, it seems incomprehensible that with forelimb structures built of the same bones a turtle and a whale should swim, a horse run, a person write and a bird or bat fly. An engineer could design better limbs for each purpose. But if we accept that these animals inherited their skeletal structures from a common ancestor and became modified only as they adapted to different ways of life, then the similarity of their structures makes sense.

Scientists call such structures homologous or inherited similarities, and have concluded that they are best explained by common descent; that is, the homologous structure evolved in an ancestor common to all the species exhibiting the homology and was consequently modified by natural selection to fit different lifestyles. Comparative anatomists investigate such homologies, not only in bone structure but also in other parts of the body, working out relationships from degrees of similarity. The correspondence of structures is typically very close among some organisms – the different varieties of songbirds, for instance – but becomes less so as the organisms being compared are less closely related in their evolutionary history. The similarities are fewer between mammals and birds than they are among mammalian species, and they are fewer still between mammals and fishes. Similarities in structure, therefore, not only manifest evolution but also help to reconstruct the phylogeny, or evolutionary history, of organisms.

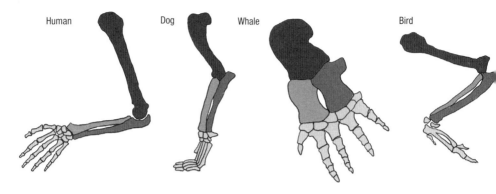

FORELIMBS OF FOUR VERTEBRATES SHOWING SIMILAR STRUCTURES ARRANGED IN SIMILAR WAYS, ALTHOUGH THEY ARE USED FOR DIFFERENT FUNCTIONS: FOR WRITING (HUMAN), RUNNING (DOG), SWIMMING (WHALE), FLYING (BIRD).

Comparative anatomy also reveals why most organismic structures are not perfect. Like the forelimbs of turtles, horses, humans, birds and bats, an organism's body parts are less than perfectly adapted because they are modified from an inherited structure rather than designed from completely raw materials for a specific purpose. The anatomy of animals shows that it has been designed to fit their lifestyles, but it is 'imperfect' design, accomplished by natural selection, rather than 'intelligent' design, as it would be if designed by an engineer. The imperfection of structures is evidence for evolution, contrary to the creationists' arguments asserting that the design of organisms demonstrates that they were fashioned by the Creator, just as the design of a watch demonstrates that it has been fashioned by a watchmaker.

Separate lineages sometimes independently evolve similar features, known as analogous structures. Although these look like homologies, they result not from common ancestry but rather from 'convergence', because the organisms live in common environments. For example, dolphins are aquatic mammals that have evolved from terrestrial mammals over the past 50 million years. In evolutionary terms, they are as distant from fish as are mice or humans. But they have evolved streamlined bodies that closely resemble the bodies of fish, sharks and even extinct dinosaurs known as ichthyosaurs. Evidence from many different fields of biology (morphology, physiology, ecology, etc.) allows evolutionary biologists to discern whether physical and behavioural similarities are the product of common descent or are independent responses to similar environmental challenges. Typically, analogous features are similar in overall configuration, but not in the details, unlike homologous structures. Thus, for example, the skeletons of forelimbs of humans, dogs, whales and birds consist of the same components arranged in similar ways: humerus, radius and ulna. The number and arrangement of carpals, metacarpals and phalanges are also the same in humans, dogs and whales, although they are

The imperfection of structures is evidence for evolution, contrary to the creationists' arguments asserting that the design of organisms demonstrates that they were fashioned by the Creator.

somewhat modified in birds. This precise correspondence in several organisms between the component parts of a limb or an organ is evidence of common descent from an ancestor that had a similar structure with the same components arranged in the same way; that is, the structures are homologous. The anatomies of a dolphin and a tuna are clearly designed for swimming and living in the oceans, but the details of the design are quite different (for example, tunas, unlike dolphins, have gills for breathing), showing that they are analogous. Similarly, the wings of both a bird and a bat are designed for flying, but they are quite different in their anatomical detail; from this we can tell that they evolved independently rather than being inherited from the same winged ancestor.

Similarity of embryos

Evolutionists, following Darwin, find evidence for evolution in the similarities in the development of embryos; that is, in the development of organisms from fertilised egg to moment of birth or hatching. Vertebrates, from fishes through lizards to humans, develop in ways that are remarkably similar during the early stages, but they become more and more differentiated as the embryo approaches maturity. The similarities persist longer between organisms that are more closely related (e.g. humans and monkeys) than between those less closely related (such as humans and lizards).

Common developmental patterns reflect evolutionary kinship. Lizards and humans share a developmental pattern inherited from their remote common ancestor; the inherited pattern of each was modified only as the separate descendant lineages evolved in different directions. The common embryonic stages of the two creatures reflect the constraints imposed by this shared inheritance, which prevents changes that have not been necessitated by their diverging environments and ways of life.

The embryos of humans and other non-aquatic vertebrates exhibit gill slits even though they never breathe through gills. These slits are found in the embryos of all vertebrates because they share as common ancestors the fish in which these structures first evolved. Human embryos also exhibit, by the fourth week of development, a well-defined tail, which reaches its maximum length at six weeks. Similar embryonic tails are found in other mammals, such as dogs, horses and monkeys; in humans, however, the tail eventually shortens, persisting only as a rudiment in the adult coccyx. Embryonic rudiments are inconsistent with claims of intelligent design: why would a structure be designed to form during early development if it will disappear before birth? Evolution makes sense of embryonic rudiments.

A close evolutionary relationship between organisms that appear drastically different as adults can sometimes be recognized through their embryonic homologies. Barnacles, for example, are sedentary crustaceans with little apparent likeness to free-swimming crustaceans such as lobsters, shrimps or copepods. Yet they pass through a free-swimming larval stage, the nauplius, which is unmistakably similar to that of other crustacean larvae.

Embryonic rudiments that never fully develop, such as the gill slits in humans, are common in all sorts of animals. Some, however, persist as adult vestiges, reflecting evolutionary ancestry. A familiar rudimentary organ in humans is the vermiform appendix. This wormlike structure attaches to a short section of intestine called the caecum, which is located at the point where the large and small intestines join. The human vermiform appendix is a functionless vestige of a fully developed organ present in mammals such as the rabbit and other herbivores, where a large caecum and appendix store vegetable cellulose to enable its digestion with the help of bacteria. Vestiges are instances of imperfections – like those seen in anatomical structures – that argue against creation by design but are fully understandable as a result of evolution by natural selection.

Biogeography

The varying geographic distribution of plants and animals throughout the world and the distinctive floras and faunas of island archipelagos were, for Darwin, evidence of evolution, eventually reinforced by later knowledge. The diversity of life is stupendous. Approximately 250,000 species of living plants, 100,000 species of fungi, and more than one million species of animals have been described and named, each occupying its own peculiar ecological setting or niche, and the census is far from complete. Some species, such as human beings and our companion the dog, can live under a wide range of environmental conditions. Others are amazingly specialized. One species of the fungus *Laboulbenia* grows exclusively on the rear portion of the covering wings of a single species of beetle (*Aphaenops cronei*) found only in some caves of southern France. The larvae of the fly *Drosophila carcinophila* can develop only in specialized grooves beneath the flaps of the third pair of oral appendages of a land crab (*Gecarcinus ruricola*) found on certain Caribbean islands.

How can we make intelligible the colossal diversity of living beings and the existence of such extraordinary, seemingly whimsical creatures as the fungus beetle and the fly described above? And why are island groups like the Galápagos so often inhabited by forms similar to those on the nearest mainland but belonging to different species? Evolution explains that biological diversity results from the descendants of local or migrant predecessors becoming adapted to their diverse environments. This explanation can be tested by examining present species and local fossils to see whether they have similar structures, which would indicate how one is derived from the other. There should also be evidence that forms without an established local ancestry have migrated into the locality.

Wherever such tests have been carried out, these conditions have been confirmed. A good example is provided by the mammalian populations of North and South America, where strikingly different endemic forms evolved in isolation until the emergence of the isthmus of Panama approximately three million years ago. Thereafter, the armadillo, porcupine and opossum – mammals of South American origin – migrated north, along with many other species of plants and animals, while the mountain lion and other North American species made their way across the isthmus to the south.

Each of the world's continents has its own distinctive collection of animals and plants. In Africa there are rhinoceroses, hippopotamuses, lions, hyenas, giraffes, zebras, lemurs, monkeys with narrow noses and non-prehensile tails, chimpanzees and gorillas. South America, which extends over much the same latitudes as Africa, has none of these animals; instead, it has pumas, jaguars, tapirs, llamas, raccoons, opossums, armadillos and monkeys with broad noses and large prehensile tails. Australia boasts a great diversity of marsupial mammals, which lack placentas, so that much of the early development takes place in a mother's external pouch rather than inside the womb. Marsupials include kangaroos but also Australian moles and anteaters and Tasmanian wolves.

The vagaries of biogeography are not due solely to the suitability of the different environments. There is no reason to believe that South American animals are not well suited to living in Africa or those of Africa to living in South America. When rabbits were intentionally introduced in Australia, so that they could be hunted for sport, they prospered beyond the expectations of the introducers and became an agricultural pest. Hawaii lacks native land mammals, but when feral pigs and goats were brought to the islands in the nineteenth century for hunting, they multiplied to large numbers and are now endangering the native vegetation.

An interesting story is the case of Santa Catalina Island, some 25 miles south-west of Los Angeles. Bison were introduced to the island in the 1940s by some filmmakers. After their filming was completed, they did not bother removing the animals. Santa Catalina Island does not have any native mammal larger than a fox. Yet the bison prospered and successfully reproduced throughout the uninhabited parts of the island, threatening the vegetation. The bison are now regularly culled so that only two herds with a few dozen animals are kept, and have become a distinctive tourist attraction.

The remarkable diversification of life in different parts of the world is evidence of evolution promoted by natural selection. Even though climate and other features of the environment may be comparable at similar latitudes, the flora and fauna are diverse on different continents and on different islands. This diversity occurs because natural selection depends on the opportunism of genetic mutations, which are random events. Moreover, evolution relies on previous changes, so that diversification from one continent or island to another, or between continents and islands, is cumulative over time. Evolutionary change occurs in response to the environment, but it is conditioned by history: mammals do not evolve into fishes, nor insects into molluscs.

> *The remarkable diversification of life in different parts of the world is evidence of evolution promoted by natural selection.*

Darwin's observations of the flora and fauna of South America, so different from those of the Old World, convinced him of the reality of evolution. The evidence from biogeography is also apparent on a scale much smaller than continental: Darwin observed that different Galápagos islands had different kinds of tortoises and different species of finches, which in turn were different from those found in continental South America. He was startled by the Galápagos' tortoises, giant lizards, mockingbirds and finches, different as they were from mainland species and diverse among the islands as well.

Hawaii's cauldron

The Galápagos islands are on the equator, about 600 miles west of the South American country of Ecuador. More remote yet than the Galápagos are the Hawaiian islands, more than 2,500 miles away from the North American mainland. Many sorts of plants and animals are lacking in Hawaii, whereas others are endemic (i.e. native nowhere else on Earth) and are extraordinarily diverse. The following table lists groups of organisms with numerous and very diverse species native to Hawaii.

	NUMBER OF SPECIES	PER CENT ENDEMIC
Ferns	168	65
Flowering plants	1,729	94
Snails	1,064	99+
Drosophila	510	100
Other insects	3,750	99+
Land mammals	0	0

Kohala, the oldest volcano on the large island of Hawaii, is somewhat less than a million years old; of the other volcanoes, Mauna Kea and Mauna Loa are much younger, and Kilauea is still active. The island of Kauai was formed less than 10 million years ago; other islands are of intermediate age, increasingly older from south-east to north-west. The gradual formation over millions of years of these volcanic islands has resulted in successive colonizations by plants and animals, and therefore species diversification. *Drosophila* fruit flies are favoured by experimental geneticists because they can easily and inexpensively be cultured in the laboratory. The ecology, behaviour and genetics of Hawaiian fruit flies have been studied intensively. There are about 1,500 known species of *Drosophila* flies in the world; nearly one third of them live in Hawaii and nowhere else, although the total area of the archipelago is less than one twentieth the area of California. Moreover, the morphological and behavioural diversity of Hawaiian *Drosophila* exceeds that of *Drosophila* in the rest of the world. There are more than 1,000 species of land snails in Hawaii, all of which have evolved in the archipelago; and about 80 bird species, all but one of which exist nowhere else.

Why has such 'explosive' evolution occurred in Hawaii? The overabundance of fruit flies there contrasts with the absence of many other native insects, such as mosquitoes and cockroaches. Because of their remote isolation, the Hawaiian islands have rarely been reached by colonizing plants and animals. Some that did reach the islands found suitable habitats without competitors or predators. The ancestors of Hawaiian fruit flies were passively transported to the archipelago by air currents or flotsam before other groups of insects reached it, and there they found a multitude of opportunities for living. They rapidly evolved and diversified by exploiting the available resources. It is known from genetic studies that several hundred species have derived from a single colonizing species; they adapted to the variety of opportunities available in diverse niches by evolving suitable adaptations, which range broadly from one species to another. In Hawaii, some *Drosophila* species feed on decaying leaves on the forest floor; others feed on flowers; still others on fungi, and so on.

The geographic remoteness of the Hawaiian islands is a more reasonable explanation for the explosive diversity of a few kinds of organisms – such as fruit flies, snails and birds – than an inordinate preference on the part of the Creator for providing the archipelago with numerous flies, or a peculiar distaste for creating mosquitoes, cockroaches and other insects there. There are no native land mammals in Hawaii; no mammals existed there until pigs and goats were introduced by humans. Hawaii was never colonized by mammals because none happened to reach the archipelago from the distant continents where mammals lived.

The Hawaiian islands are no less hospitable than other parts of the world. The absence of many kinds of organisms, and the great multiplication of a few kinds, is due to the fact that many sorts of organisms never reached the islands because of their geographic isolation. Those that did diversified over time because of the absence of related organisms that would compete for resources.

WHAT IS NATURAL SELECTION?
Design without designer

'Can we doubt … that individuals having any advantage, however slight, over others, would have the best chance of surviving and of procreating their kind? On the other hand, we may feel sure that any variation in the least degree injurious would be rigidly destroyed. This preservation of favourable variations and the rejection of injurious variation, I call Natural Selection.'[2]

In the sentence preceding the quotation above, Darwin rhetorically asks: 'Can it, then, be thought improbable, seeing that variations useful to man have undoubtedly occurred, that other variations useful in some ways to each being in the great and complex battle of life, should sometimes occur in the course of thousands of generations?' He is referring to Chapter 1 of *On the Origin of Species*, entitled 'Variation under Domestication'. Darwin collected every breed of domestic pigeon that he could find. 'The diversity of the breeds is something astonishing,' he writes, and dedicates nine pages to describing a number of them. More generally, he is saying that the experience of agriculture has taught us that animals and plants from time to time show new variants in their traits, so that farmers can select desirable features; say, corn with larger kernels or cows that produce more milk. These variants are heritable, that is, they are transmitted to the offspring. If variations useful to man have occurred, it must be the case, says Darwin, that variants that are beneficial to the organisms themselves must also occur from time to time, such as increased running speed in a cheetah or better dispersal of seed in an oak. These variants are 'useful' to the organisms precisely because they increase their chances of survival and procreation. This in turn means that these advantageous variations will be multiplied over the generations at the expense of less advantageous variants. This is the process known as 'natural selection'. The key point is that natural selection – the 'preservation of favourable variations and the rejection of injurious variations', in Darwin's words – accounts for the 'design' of organisms: why they are constructed so that they can function in the environments in which they live. A fleet cheetah captures more prey; a tree with more leaves captures sunlight more effectively.

Natural selection

Darwin's ground-breaking theory of natural selection explains the incremental process that has been occurring for millions and millions of years, since there was life – that is, organisms that reproduce – on Earth. Natural selection accounts for evolution because adaptive changes accumulate or replace less adaptive ones over aeons. It accounts for the diversification of species, because different variants may be more or less advantageous at different times or in different places. Most fundamentally, it accounts for the

Natural selection is an incremental process that has been occurring for millions and millions of years.

'design' of organisms and their features. The complex 'camera-eye' of humans or octopuses came about because since time immemorial, vision was beneficial to their ancestors, so variants that improved the perception of light gradually accumulated.

Darwin dedicates a large part of *Origin of Species* to explaining how natural selection works. Much more has been learned by scientists in the past century and a half. If a short definition that catches the core of the process is desired, we can say that natural selection is 'the differential reproduction of hereditary variations', which is how textbooks often define it. That is saying simply that useful variants multiply more effectively over the generations than less useful (or harmful) variants. Thus a cheetah able to run faster will catch more prey, and therefore live longer and leave more offspring than a slower cheetah. So, a hereditary variant that boosts fleetness will increase in frequency over the generations and eventually replace the slower variant.

The brief definition given in the preceding paragraph does not, however, provide a satisfactory understanding of the process of natural selection and how it accounts for the evolution of organisms and their design. In the same way, we would not learn much about the Earth by defining it as 'the third planet that revolves around the Sun'. We can increase our understanding by extending the definition as follows: 'Natural selection is the differential reproduction of alternative variations, determined by the fact that some variations are beneficial because they increase the probability that the organisms having them will live longer or be more fertile than organisms having alternative variations.' This definition adds the reason why differential reproduction occurs, namely because some variants are more useful than others. It also specifies the two main components of reproduction: survival and fertility. A definition even more informative could refer to the outcome of the process, adding to the previous definition: 'Over the generations, beneficial variations will be preserved and multiplied; injurious or less beneficial variations will be eliminated.' We could also refer in the definition to the long-term consequences of the process: 'Over long periods of time, natural selection usually changes the make-up and functioning of organisms and causes their diversification (i.e. multiplication of species) as they adapt to different environments.'

There are many books and innumerable scientific papers that expound on the complexities of the process of natural selection; publications in which appropriate mathematical models and equations are developed that account for the process of natural selection; and others that report laboratory experiments or investigations of natural selection in nature. In addition to *Origin of Species*, Darwin wrote several books further describing how natural selection works; books dedicated, for example, to the evolution of orchids, barnacles, earthworms, primates and humans.

Heredity and mutation

Hereditary variations, whether favourable or not to the organisms, arise by a process known as mutation, which changes one gene into another. For example, a gene that causes a plant to be short mutates into a gene that causes a plant to be tall. The details of the mutation process were gradually elucidated in the twentieth century. With regard to evolution, what is significant is that unfavourable mutations are eliminated by natural selection because their carriers leave fewer descendants than those carrying alternative favourable mutations, which, in turn accumulate over the generations. Different mutations are favoured in different environments, or habitats; as the habitats change (or organisms colonize new ones), the organisms will evolve.

Mutation is said to be a random process. What is meant is that mutations arise without regard to their effects on the organisms' ability to survive and reproduce. If mutation were the only process affecting evolutionary change, the organization of living things would gradually disintegrate. Natural selection keeps the disorganizing effects of mutation in check because it multiplies beneficial mutations and eliminates harmful ones. An analogy would be an editor who would pick up the small proportion of letters or words that improve the meaning of a text and discard nonsensical changes.

Natural selection accounts not only for the preservation and improvement of the organization of living beings, but also for their diversity.

As noted above, natural selection accounts not only for the preservation and improvement of the organization of living beings, but also for their diversity. In different localities or different circumstances, natural selection favours different traits, precisely those that make the organisms well adapted to the particular circumstances and ways of life.

Mutation rates have been measured in a great variety of organisms, mostly for mutants that exhibit conspicuous effects. In humans and other multicellular organisms, the rate is typically around 1 mutation per 1,000,000 sex cells. For example, a new mutation that causes sickle-cell anaemia might appear in one of every 500,000 births. (Each individual results from two sex cells, the ovum and the sperm.) Although mutation rates are low, new mutants appear continuously in nature because there are many individuals in every species and many genes in every individual. The human population consists of about 7 billion people. If any given mutation occurs once for every 500,000 people, living humans would collectively carry 14,000 copies of every possible mutation.

The process of mutation provides each generation with new genetic variations, in addition to those carried over from previous generations. Thus, it is not surprising to see that, when new environmental challenges arise, species are able to adapt to them. More than 100 insect species, for example, have developed resistance to the pesticide DDT in parts of the world where spraying has been intense. Although these insects had never before encountered this synthetic compound, mutations occurred that allowed them to survive in its presence. The adaptation (resistance to DDT) was rapidly multiplied by natural selection.

The resistance of disease-causing bacteria and parasites to antibiotics and other drugs is a consequence of the same process. When an individual receives an antibiotic that specifically kills the bacteria causing a disease – say, tuberculosis – the immense majority of the bacteria die, but one in several million of them may have a mutation that provides resistance to the antibiotic. These resistant bacteria will survive and multiply, and eventually that antibiotic will no longer cure the disease. This is why modern medicine treats bacterial diseases with cocktails of antibiotics. If the incidence of a mutation conferring resistance for a given antibiotic is one in a million, the incidence of one bacterium carrying three mutations, each conferring resistance to one of three antibiotics, is one in a quintillion (one in a million million million); it is not likely, if not altogether impossible, that such bacteria will exist in any infected individual.

A creative process

Natural selection is a creative process. Although it does not 'create' the component entities upon which it operates (genes and genetic mutations), it does yield 'design', combinations that could not have existed otherwise and that are beneficial to the organisms.

Proponents of intelligent design and other creationists argue against evolution by assuming that it is a random process. However, the combination of genetic units that carry the hereditary information responsible for the formation of the vertebrate eye, for example, could never have been produced purely by chance, not even if we allow for the three-billion-plus years during which life has existed on Earth. The complicated anatomy of the eye, like the exact functioning of the kidney, is the result of a non-random process – natural selection.

An experiment carried out with *Escherichia coli*, single-celled bacteria that occur in the colon of humans and other mammals, may illustrate how natural selection accumulates beneficial hereditary variants. Some strains of *E. coli,* in order to reproduce in a culture – a small test tube with a water solution of sugar – require a certain substance, the amino acid histidine, to be provided with the sugar. When a few such bacteria are added to a small test tube with a culture solution that includes histidine, they multiply rapidly and produce between 20 and 30 billion bacteria in one or two days. If a drop of the antibiotic streptomycin is added to the culture, most bacteria will die, but after a day or two the culture will again teem with billions of bacteria. How come? Spontaneous genetic mutations causing resistance to streptomycin occur in normal (i.e. non-resistant) bacteria randomly, at rates in the order of 1 in 100 million bacterial cells. In a bacterial culture with 20 to 30 billion bacteria, we expect between 200 and 300 bacteria to be resistant. When streptomycin is added to the culture, only the resistant cells survive. The 200 or 300 surviving bacteria will reproduce, and allowing one or two days for the necessary number of cell divisions, 20 billion or so bacteria are produced, all resistant to streptomycin.

Consider now a second step in this experiment. The streptomycin-resistant cells are transferred to a culture with streptomycin but without histidine (the amino acid that they require in order to grow and reproduce). Most of the bacteria will fail to reproduce and will die; yet after a day or two, the culture will be teeming with billions of bacteria. This is because among cells that need the amino acid histidine to grow, mutants able to reproduce in the absence of histidine arise spontaneously at rates of about 4 in 100 million bacteria. If the culture has 20 to 30 billion bacteria, about 1,000 will survive in the absence of histidine and will reproduce until the available medium is saturated with them.

Natural selection has produced, in two steps, bacterial cells resistant to streptomycin and not requiring histidine for growth. The probability of these two mutations happening in the same bacterium is about 4 in 10 million billion cells. An event of such low probability is unlikely to occur even in a large laboratory culture of bacterial cells. Yet natural selection commonly results in cells possessing both properties. A complex trait made up of two components has come about in real time by natural processes. It can readily be understood that the example can be extended to three, four and more component steps. At the end of a long process of evolution, we have organisms each exhibiting features 'designed' for survival in its habitat.

No chance

The fact that evolution is not the outcome of random processes deserves further emphasis. There is a selection process that picks up adaptive combinations because these reproduce more effectively and thus come to prevail in populations. Simple adaptive combinations constitute, in turn, new levels of organization upon which the mutation (random) plus selection (non-random or directional) processes again operate. The organizational complexity of animals and plants has arisen as a consequence of natural selection acting one step at a time, over aeons of time.

Several hundred million generations separate modern animals from the early animals of the Cambrian geological period (542 million years ago). The number of mutations that can be tested, and those eventually selected, in millions of individual animals over millions of generations is difficult for the human mind to fathom, but what we *can* understand is that the accumulation of thousands or millions of small, functionally advantageous changes could yield remarkably complex and adaptive organs (see *Is Creationism True?*).

How complex organs, such as so-called 'camera' eyes, may arise stepwise through organs of intermediate complexity is manifest in living molluscs (squids, clams and snails), where a gradation can be found from a very simple eye (an eye spot consisting of a few pigmented cells with nerve fibres attached to them, as found in limpets), through a pigment cup (slit-shell molluscs), to an optic cup with a pinhole serving the role of lens (open-ocean *Nautilus*), to an eye with a primitive refractive lens protected by a layer of skin cells serving as a cornea (*Murex*

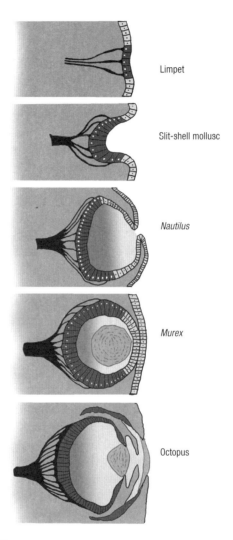

Limpet

Slit-shell mollusc

Nautilus

Murex

Octopus

EYES IN LIVING MOLLUSCS FROM VERY SIMPLE, AS IN THE LIMPET *PATELLA* (TOP) TO INCREASINGLY MORE COMPLEX, AS IN THE SLIT-SHELL *PLEUROTOMARIA*, THE MARINE SNAILS *NAUTILUS* AND *MUREX*, AND THE COMPLEX CAMERA-EYE OF THE OCTOPUS (BOTTOM).

marine snails), to the eye of octopuses and squids, as complex as the human eye, with cornea, iris, refractive lens, retina, vitreous internal substance, optic nerve and muscle.

Selection in action

Evolution by natural selection is an incremental process, operating over time and yielding organisms better able to survive and reproduce than others, which typically differ from one another at any one time only in small ways; for example, the difference between having or lacking an enzyme able to synthesize the amino acid histidine. Numerous adaptations are known that involve one or only a few genes, as in the bacterial example described earlier. Another example occurs in some pocket mice (*Chaetodipus intermedius*) that live in rocky outcrops in Arizona.

Light, sandy-coloured mice are found in light-coloured habitats, whereas dark (melanic) mice prevail among dark rocks formed from ancient flows of basaltic lava. The match between background and fur colour protects the mice from avian and mammal predators guided largely by vision. Mutations in one single gene (coding for the melanocortin 1 receptor, represented as MC1R) account for the difference between light and dark peltage.

> *The organizational complexity of animals and plants has arisen as a consequence of natural selection acting one step at a time, over aeons of time.*

Adaptations relating to complex structures, functions or behaviours typically involve numerous genes. Many familiar mammals, though not marsupials, have a placenta. Marsupials include the kangaroo and other mammals native primarily to Australia and South America. Dogs, cats, mice, donkeys and primates are placental. The placenta makes it possible to extend the time the developing embryo is kept inside the mother, thus ensuring that the newborn is better prepared for independent survival. However, the placenta requires complex adaptations, such as the suppression of harmful immune interactions between mother and embryo, delivery of suitable nutrients and oxygen to the embryo and the disposal of embryonic wastes.

The mammalian placenta evolved more than 100 million years ago and proved to be a successful adaptation, leading to the extinction of most marsupial species in the continents of the Old

World and North America. The placenta has also evolved in some fish groups, such as *Poeciliopsis*. In some species, the females supply the yolk in the egg, which furnishes nutrients to the developing embryo, but do not directly contribute nutrients. Other *Poeciliopsis* species, however, have evolved a placenta through which the mother provides additional nutrients to the developing embryo. The reconstruction of the evolutionary history of *Poeciliopsis* species, by means of molecular biology, has shown that the placenta evolved independently three times in this fish group and that the required complex adaptations accumulated in less than 750,000 years.

Increased complexity is not a necessary consequence of natural selection, but it does emerge occasionally. From time to time, a mutation that increases complexity will be favoured by natural selection over mutations that do not, such as occurred in the evolution of the eye in molluscs and vertebrates. Complexity-increasing mutations, however, do not necessarily accumulate over time. The longest-living groups of organisms on Earth are the microscopic bacteria, which have existed continuously on our planet for 3.5 billion years or so and yet exhibit no greater complexity than their ancient ancestors. More complex organisms came about later, without the elimination of their simpler relatives. Over the aeons, multitudes of complex organisms have arisen on Earth. For example, the primates appeared only 50 million years ago; our species, *Homo sapiens*, less than 200,000 years ago.

Natural selection is not by itself a creative process because it does not create the raw materials, the randomly occurring mutations. However, it *becomes* a creative process, causing favourable mutations to spread over multiple generations to the whole species, and accumulating different mutations favourable to organisms.

WHAT IS SURVIVAL OF
THE FITTEST?
The evolutionary synthesis

'*Variations, however slight . . . if they be in any degree profitable
to the individual of a species . . . will tend to the preservation
of such individuals, and will generally be inherited by the offspring . . .
I have called this principle, by which each slight variation, if useful,
is preserved, by the term of Natural Selection . . . But the expression
often used by Mr Herbert Spencer of the Survival of the Fittest is
more accurate, and is sometimes equally convenient.*'[3]

'The survival of the fittest' is, as Darwin saw it, another expression
for natural selection. The phrase 'natural selection' points to the
process of differential survival and reproduction, which accounts
for the adaptation of organisms to the environmental conditions
of life. The survival of the fittest focuses instead on the *outcome* of
that process. Those that are most likely to survive and reproduce
are precisely those with more profitable variations, those that are
most fit, the fittest.

Darwin did not coin 'the survival of the fittest'. The
author of that famous phrase was his younger contemporary, the
philosopher Herbert Spencer. Darwin introduced the phrase for
the first time in the fifth edition of *On the Origin of Species*, which
was published in 1869. (A measure of the book's swift public
success can be glimpsed by considering its rapid succession of
editions and the number of copies printed. The publisher, John
Murray of London, had been somewhat apprehensive about the

venture. Eventually he agreed to print 1,250 copies, which, as is well known, were sold, largely by subscription, on the first day they were offered for sale. The second edition was published three months later, in December 1859, with a print run of 3,000 copies; the third in 1861, with 2,000 copies; the fourth in 1866, with 1,500 copies; the fifth in 1869, with 2,000 copies; the sixth edition, the last one published during Darwin's lifetime, was published in 1872, with a print run of 3,000 copies.)

Malthus, Wallace and Darwin

Darwin credited the Rev. Thomas Malthus as the source of inspiration for the idea of natural selection as the process that would explain evolution and adaptation. In 1798 Malthus had published a short essay, only 600 words long, on the subject of human population. He wrote:

> *Population, when unchecked, increases in a geometrical ratio. Subsistence increases only in an arithmetical ratio. A slight acquaintance with numbers will shew the immensity of the first power in comparison of the second … This implies a strong and constantly operating check on population from the difficulty of subsistence.*

This is how Darwin, in his autobiography, describes his acquaintance with Malthus' essay:

> *In October 1838, that is, fifteen months after I had begun my systematic enquiry* [and barely two years after returning from his five-year trip around the world in the HMS *Beagle*], *I happened to read for amusement 'Malthus on Population' and being well prepared to appreciate the struggle for existence which everywhere goes on from long-continued observation of the habits of animals and plants, it at once struck me that under these circumstances favourable variations would tend to be preserved, and unfavourable ones to be destroyed. The result of this would be the formation of new species. Here then I had at last got a theory by which to work.*

In 1858 Alfred Russel Wallace, a naturalist who was at the time collecting insects, birds, and mammals in the islands of South East Asia, came upon the idea of natural selection independently of Darwin (see also *What is Evolution?*). Remarkably, Wallace's source of inspiration was also Malthus. As he wrote in 1891:

> *At the time* [February 1858] *I was suffering from a rather severe attack of intermittent fever at Ternate in the Moluccas . . . and something led me to think of the positive checks described by Malthus in his 'Essay on Population', a work I had read several years before, and which had made a deep and permanent impression on my mind. These checks – war, disease, famine, and the like – must, it occurred to me, act on animals as well as on man. Then I thought of the enormously rapid multiplication of animals, causing these checks to be much more effective in them than in the case of man; and, while pondering vaguely on this fact there suddenly flashed upon me the idea of the survival of the fittest – that the individuals removed by these checks must be on the whole inferior to those that survived.*

There had been other independent discoveries, or at least inklings, of the idea of natural selection, notably by the American physician turned physiologist William Wells. Darwin learned about Wells' ideas after publication of *On the Origin of Species*, and added a reference to him in its fourth edition, stating that Wells was the first author who 'distinctly' recognized 'the principles of natural selection'.

The evolutionary synthesis: natural selection and genetics

The most serious difficulty facing Darwin's evolutionary ideas was the lack of an adequate theory of inheritance that would account for the preservation through the generations of the variations on which natural selection was supposed to act.

The missing link in Darwin's argument was provided by Mendelian genetics. Gregor Mendel's discoveries, published in 1866, remained unknown to Darwin, and, indeed, did not become generally known by the scientific community until 1900, when they were independently rediscovered by several scientists on the continent (see *What are Chromosomes, Genes and DNA?*). A suitable integration of genetics and natural selection did not happen until the 1920s and 1930s, when several scientists used mathematical arguments to show that small, seemingly continuous variations (in such characteristics as body size, number of eggs laid and the like) could be explained by Mendel's laws. Distinguished members of this group of theoretical geneticists were R.A. Fisher and J.B.S. Haldane in Great Britain and Sewall Wright in the United States. These and other scientists also demonstrated that natural selection acting cumulatively on small variations could yield over time major evolutionary changes in form and function. Their work provided a theoretical framework for the integration of genetics into Darwin's theory of natural selection, but it had a negligible impact on contemporary biologists because it was formulated in a mathematical language that most biologists could not understand, and was limited in scope, omitting many issues, such as speciation (the process by which new species are formed), that were of great importance to biologists.

A major breakthrough came in 1937 with the publication of *Genetics and the Origin of Species* by Theodosius Dobzhansky, a Russian-born American naturalist and experimental geneticist. Dobzhansky's book advanced a reasonably comprehensive account of the evolutionary process in genetic terms, laced with experimental evidence supporting the theoretical argument. *Genetics and the Origin of Species* may be considered the most important landmark in the formulation of what came to be known as the 'synthetic theory of evolution', or the 'evolutionary synthesis', effectively combining Darwinian natural selection and Mendelian genetics. It had an enormous impact on naturalists and experimental biologists, who rapidly embraced the new understanding of the evolutionary process as one of genetic change in populations. Interest in evolutionary studies was greatly stimulated, and contributions to the theory soon began to follow, extending the synthesis of genetics and natural selection to a variety of biological fields.

Other writers who may be considered major early contributors to the synthetic theory were the German-born American zoologist Ernst Mayr, the English zoologist Julian Huxley, the American palaeontologist George Gaylord Simpson and the American botanist George Ledyard Stebbins. These researchers contributed to a burst of evolutionary studies in both the traditional biological disciplines and some emerging ones – notably population genetics and, later, evolutionary ecology. By 1950, acceptance of Darwin's theory of evolution by natural selection was universal among biologists, and the synthetic theory had become widely adopted.

The unfit, the fit, the fittest

Natural selection refers to any reproductive bias favouring some genes or genotypes over others. It promotes the adaptation of organisms to the environments in which they live; any hereditary variant that improves the ability to survive and reproduce in an environment will increase in frequency over the generations, precisely because the organisms carrying such a variant will leave more descendants than those lacking it. Hereditary variants arise by mutation. Unfavourable ones are eventually eliminated by natural selection; their carriers leave fewer descendants than those carrying alternative favourable variants. Favourable mutations increase in frequency over the generations. The process continues indefinitely because the environments that organisms inhabit are forever changing.

Natural selection refers to any reproductive bias favouring some genes or genotypes over others.

Environments change physically – in their climate, configuration and so on – but also biologically, because the predators, parasites, competitors and food sources with which an organism interacts are themselves evolving.

The parameter used to measure the effects of natural selection is fitness, which can be expressed as an absolute or as a relative value. Consider a population having at a certain gene locus two alleles, A_1 and A_2, and hence three genotypes: A_1A_1, A_1A_2 and A_2A_2. Assume that on average each A_1A_1 and each A_1A_2 individual produces one offspring but that each $A_2 A_2$ individual produces two. One could use the average number of progeny left by each genotype as a measure of that genotype's absolute fitness and calculate the changes in gene frequency that would occur over the generations, if the actual number of progeny produced by each of the three genotypes was known. Evolutionists, however, find it mathematically more convenient in most calculations to use relative fitness values, which they represent with the letter w. They usually assign the value 1 to the genotype with the highest reproductive efficiency and calculate the other relative fitness values proportionally. For the example just used, the relative fitness of the $A_2 A_2$ genotype would be $w = 1$ and that of each of the other two genotypes would be $w = 0.5$. A parameter related to fitness is the selection coefficient, often represented by the letter s, which is defined as $s = 1 - w$. The selection coefficient is a measure of the reduction in fitness of a genotype. In the example above, the selection coefficients are $s = 0$ for $A_2 A_2$ and $s = 0.5$ for $A_1 A_1$ and $A_1 A_2$.

Selection of single genes

The different ways in which natural selection affects gene frequencies are illustrated by the following examples. Suppose now that one homozygous genotype, $A_2 A_2$, has lower fitness than the other two genotypes, $A_1 A_1$ and $A_1 A_2$. (This is the situation in many human diseases, such as phenylketonuria – PKU – and sickle-cell anaemia, that require the presence of two deleterious mutant alleles for the trait to manifest.) The heterozygotes and the homozygotes for the normal allele (A_1) have equal fitness, higher than that of the homozygotes for the deleterious mutant allele (A_2). Call the fitness of these latter homozygotes $1-s$ (the fitness of the other two genotypes is 1), and let p be the frequency of A_1 and q the frequency of A_2. It can be shown that the frequency of A_2 will decrease each generation by an amount given by $\Delta q = -spq^2 / (1 - sq^2)$. The deleterious allele will continuously decrease in frequency until it has been eliminated. The rate of elimination is fastest when $s = 1$ (i.e. when the relative fitness $w = 0$); this occurs with fatal diseases, such as untreated PKU, when the homozygotes die before the age of reproduction.

Because of new mutations, the elimination of a deleterious allele is never complete. A dynamic equilibrium frequency will exist when the number of new alleles produced by mutation is the same as the number eliminated by selection.

Overdominance

In many instances heterozygotes have a higher degree of fitness than homozygotes for one or the other allele. This situation, known as heterosis or overdominance, leads to the stable coexistence of both alleles in the population and hence contributes to the widespread genetic variation found in populations of most organisms. The model situation is:

Genotype	A_1A_1	A_1A_2	A_2A_2
Fitness	$1-s$	1	$1-t$

A particularly interesting example of heterosis in humans is provided by the gene responsible for sickle-cell anaemia in world regions suffering from severe malaria. Human haemoglobin in adults is for the most part haemoglobin A, a four-component molecule consisting of two α and two β haemoglobin chains. The gene Hb^A codes for the normal β haemoglobin chain, which consists of 146 amino acids. A mutant allele of this gene, Hb^S, causes the β chain to have in the sixth position the amino acid valine instead of glutamic acid. This seemingly minor substitution modifies the properties of haemoglobin so that homozygotes with the mutant allele, $Hb^S Hb^S$, suffer from a severe form of anaemia that in most cases leads, if untreated, to death before the age of reproduction.

The Hb^S allele occurs with a high frequency in some African and Asian populations. This once puzzled scientists because the severity of the anaemia, representing a strong natural selection against homozygotes, should have eliminated the defective allele. But researchers noticed that the Hb^S allele occurred at high frequency precisely in regions of the world where a particularly severe form of malaria, caused by the parasite *Plasmodium falciparum*, was endemic. In malaria-infested regions the heterozygotes survived better than either of the homozygotes, which were more likely to die from either malaria ($Hb^A Hb^A$ homozygotes) or anaemia ($Hb^S Hb^S$ homozygotes). This hypothesis has been confirmed in various ways. In a study of 100 children who died from malaria, only 1 was found to be a heterozygote, whereas 22 were expected to be so according to the frequency of the Hb^S allele in the population.

Frequency-dependent fitness

The fitness of genotypes can vary according to their relative numbers, and genotype frequencies may change as a consequence. This is known as frequency-dependent selection. Particularly interesting is the situation in which genotypic fitnesses are inversely related to their frequencies. Frequency-dependent selection may arise because the environment is heterogeneous and because different genotypes can better exploit different sub-environments. When a genotype is rare, the sub-environments that it exploits better will be relatively abundant. But as the genotype becomes common, its favoured sub-environment becomes saturated. That genotype must then compete for resources in sub-environments that are optimal for other genotypes. It follows then that a mixture of genotypes exploits the environmental resources better than a single genotype. Plant breeders know that mixed plantings (a mixture of different strains) are more productive than single stands (plantings of one strain only), although farmers typically avoid mixed plantings for reasons such as increased sowing and harvesting costs.

Sexual preferences can also lead to frequency-dependent selection. It has been demonstrated in some insects, birds, mammals and other organisms that the mates preferred are precisely those that are rare. More generally, frequency-dependent selection may account for the nearly equal number of males and females in many sexually reproducing species. In a population with more males than females, females are favoured by natural selection. Every child must have a male and a female parent. If there are more males than females, on average each female would have a larger number of progeny. Natural selection would favour genes that reduce the frequency of males until they are not more numerous than females. In a population with more females than males, the opposite will happen: natural selection will favour genes that increase the proportion of males in the population until both sexes are numerically equal.

Frequency-dependent selection occurs in Batesian mimicry, common in butterflies and other insects. Mimicry refers to the resemblance of one species, the mimic, to another species, the model. Batesian mimicry occurs when the mimic is a palatable species to birds and other predators, but the model is not because it is distasteful or provokes vomiting or some other negative reaction. Predators that have experienced the model will also avoid mimics that look like the model. This protection diminishes as the mimics increase in frequency, so that predators would mostly experience palatable mimics rather than distasteful models. In East Africa, females of the swallowtail butterfly *Papilio dardanus* mimic several model species that are quite different from one another in appearance. The advantage of one mimic over another is determined by the abundance of the different models.

IS EVOLUTION A RANDOM PROCESS?

Chance and necessity

*O*rganisms are functionally organized, adapted to certain ways
of life and their parts designed to perform certain functions.
*Fish are adapted to live in water, kidneys are intended to regulate
the composition of blood, the human hand is made for grasping and
the eye for seeing. Adaptation and design cannot come about by
chance. It was Darwin's genius to provide a natural explanation of
the design of organisms. The seemingly purposeful aspects of living
beings could now be explained by the methods of science, as the
result of natural laws manifested in natural processes.*

The English theologian William Paley, in his *Natural Theology*
(1802), pointed out that organisms display purposefulness in the
design of their limbs and organs, such as the eye, the kidney and
the bladder, which jointly adapt each organism to its distinctive
way of life. The organized complexity of living organisms
manifests, wrote Paley, that they have come about by intentional
design, not by a random process: 'a wen, a wart, a mole, a pimple'
could come about by chance, but never an eye; 'a clod, a pebble,
a liquid drop might be', but never a watch or a telescope.
Darwin accepted that organisms and their parts are designed
for certain functions and ways of life, but went on to provide a
scientific explanation of that design. He showed that the complex
organization and functionality of living beings can be explained as
the result of a natural process – natural selection.

Genes and evolution

The central argument of Darwin's theory of evolution starts with the existence of hereditary variation. Experience with animal and plant breeding demonstrates that variations can be developed that are 'useful to man'. So, reasoned Darwin, variations must occur in nature that are favourable or useful in some way to the organism itself. Advantageous variations are preserved and multiplied from generation to generation at the expense of less advantageous ones. This is the process known as natural selection. Evolution occurs as a consequence.

Evolution can be seen as a two-step process: one step is random, the second is non-random.

Biological evolution is the process of change and diversification of living things over time, and it affects all aspects of their lives – morphology, physiology, behaviour and ecology. Underlying these changes are alterations in the hereditary materials. Hence, in genetic terms, evolution consists of changes in the organism's hereditary make-up.

Natural selection has, accordingly, been defined as the differential reproduction of alternative hereditary variants, determined by the fact that some variants increase the likelihood that the organisms having them will survive and reproduce more successfully than will organisms carrying alternative variants. Selection may be due to differences in survival, in fertility, in rate of development, in mating success or in any other aspect of the life cycle. All of these differences can be incorporated under the term 'differential reproduction' because all result in natural selection to the extent that they affect the organism's number of progeny.

Evolution can be seen as a two-step process: one step is random, the second is non-random. First, hereditary variation arises; second, selection occurs of those genetic variants that will be passed on most effectively to the following generations. Hereditary variation entails two mechanisms that are random: the spontaneous mutation of one variant to another, and the recombination of those variants through the sexual process. The transmission of the hereditary variations from one generation to another is largely governed by the non-random process of natural selection, although there is also a random component, which geneticists call 'genetic drift': some alternative genetic variants are adaptively equivalent; they neither improve nor reduce the adaptation of an organism in the environment where it lives. These adaptively neutral variants are transmitted with equal probability to the next generation, but may increase or decrease in frequency by chance.

The gene pool

The sum total of all of the genes and combinations of genes that occur in all organisms of the same species is referred to as the 'gene pool' of a species. The necessity of hereditary variation in order for evolutionary change to occur can be understood in terms of the gene pool. Assume, for instance, that in a certain population, at the gene locus that codes for the human MN blood group, there is no variation; only the M gene exists in all individuals. Evolution of the MN blood group cannot take place in such a population, since the allelic frequencies have no opportunity to change from generation to generation. In populations in which both genes M and N, are present, evolutionary change is possible.

The more genetic variation that exists in a population, the greater the opportunity for evolution to occur. As the number of variable genes increases and the number of alternative forms (alleles) of each gene becomes greater, so the likelihood grows that some alleles will change in frequency at the expense of their alternates.

The more genetic variation that exists in a population, the greater the opportunity for evolution to occur.

The British geneticist R.A. Fisher mathematically demonstrated a direct correlation between the amount of genetic variation in a population and the rate of evolutionary change by natural selection. This demonstration is embodied in his fundamental theorem of natural selection: 'The rate of increase in fitness of any organism at any time is equal to its genetic variance in fitness at that time.' This theorem has been confirmed experimentally.

The enormous reservoir of genetic variation in natural populations provides virtually unlimited opportunities for evolutionary change.

Because a population's potential for evolving is determined by its genetic variation, evolutionists are interested in discovering the extent of such variation in natural populations. Typically, insects and other invertebrates are more varied genetically than mammals and other vertebrates; and plants bred by outcrossing exhibit more variation than those bred by self-pollination. But the amount of genetic variation is in any case astounding. In most organisms, every individual represents a unique genetic configuration that will never be repeated again. The enormous reservoir of genetic variation in natural populations provides virtually unlimited opportunities for evolutionary change in response to the environmental constraints and the needs of the organisms.

Mutation and randomness

There are nearly two million known species, which are widely diverse in size, shape and way of life, as well as in the DNA sequences that contain their genetic information. What is it that has produced the pervasive genetic variation within natural populations and the genetic differences among species?

The information encoded in the nucleotide sequence of DNA is, as a rule, faithfully reproduced during replication, so that each replication results in two DNA molecules that are identical to each other and to the parent molecule. But occasionally 'mistakes', or mutations, occur in the DNA molecule during replication, so that daughter cells differ from the parent cells in at least one of the letters in the DNA sequence. A mutation first appears on a single cell of an organism, but it is passed on to all cells descended from the first. Mutations can be classified into two categories: gene, or point, mutations, which affect only one or a few letters (nucleotides) within a gene; and chromosomal mutations, which change either the number of chromosomes or the number or arrangement of genes on a chromosome. Chromosomes are the elongated structures that store the DNA of each cell.

Chance is an integral component of the evolutionary process, because the mutations that yield the hereditary variations available to natural selection arise at random. The process of mutation is random in three respects. First, spontaneous mutations are random because they are rare exceptions to the fidelity of the process of DNA replication that arise as a result of physical causes that are not deterministic (or at least cannot be determined by current scientific knowledge). Second, they are random events in the sense that there is no way of determining which gene will mutate in a particular cell or a particular individual. Third, and most important in the context of evolution, they are unoriented with respect to adaptation; they occur independently of whether or not they are beneficial or harmful to the organism. Some are beneficial, most are not. Beneficial mutations are likely to be favourably transmitted to the following generations, precisely because they improve the organism's chances of survival and reproduction. Harmful mutations are likely to be eliminated or kept in check by natural selection.

The adaptive randomness of the mutation process is counteracted by natural selection, which preserves what is useful and eliminates what is harmful. Without hereditary mutations, evolution could not happen, because there would be no variations that could be differentially conveyed from one generation to the next. But without natural selection, the mutation process would yield disorganization and extinction, because most mutations are disadvantageous. Mutation and selection have jointly driven the marvellous process that, starting from microscopic organisms, has yielded orchids, birds and humans.

Mutation rates have been measured in a great variety of organisms, mostly for mutants that exhibit conspicuous effects. In humans and other animals and plants, the rate for any given mutation in a particular gene typically ranges around 1 mutation per 1,000,000 sex cells. Although mutation rates are low, new

> *Mutation and selection have jointly driven the marvellous process that, starting from microscopic organisms, has yielded orchids, birds and humans.*

mutants appear continuously in nature, because there are many individuals in every species and many genes in every individual. Thus, the process of mutation provides each generation with many new genetic variations. Moreover, some mutations that happened in the past accumulate over the generations. It is therefore not surprising to see that when new environmental challenges arise, species are able to adapt to them. For example, numerous insect species have developed resistance to DDT and other pesticides; and many species of moths and butterflies in industrialized regions have shown an increase in the frequency of individuals with dark wings, in response to environmental pollution, an adaptation known as industrial melanism. The examples can be multiplied at will.

New mutations change gene frequencies very slowly, because mutation rates are low compared to the efficiency of natural selection. Natural selection can eliminate or propagate mutated genes very fast, in one or very few generations.

There is another random process in evolution, already mentioned above. Gene frequencies can change from one generation to another by a process of chance known as genetic drift. This occurs because populations are finite in number, and thus the frequency of a gene may change in the following generation by accidents of sampling, just as it is possible to get more or less than 50 heads in 100 throws of a coin simply by chance. The magnitude of the gene-frequency changes due to genetic drift is inversely related to the size of the population; the larger the number of reproducing individuals, the smaller the effects of genetic drift. Genetic drift can have important evolutionary consequences when a new population becomes established by only a few individuals, as, for example, in the colonization of islands and lakes.

Randomness and determinism

If mutation and drift were the only processes of evolutionary change, the organization of living things would gradually disintegrate, because they are random processes with respect to adaptation. Mutation and drift occur without regard for the consequences the changes may have in the ability of the organisms to survive and reproduce. The effects of such processes alone would be analogous to those of a mechanic who changed parts in a car engine at random, with no regard for the role of the parts. Natural selection keeps the disorganizing effects of mutation and other processes in check, because it multiplies beneficial mutations and eliminates harmful ones, accounting not only for the preservation and improvement of the organization of living beings, but also for their diversity.

The effects of natural selection can be studied by measuring the ensuing changes in gene frequencies; but they can also be explored by examining changes in the observable characteristics – or phenotypes – of individuals in a population. Distribution scales of phenotypic traits such as height, weight, number of progeny and longevity typically show greater numbers of individuals with intermediate values and fewer and fewer toward the extremes (the so-called normal distribution). When individuals with intermediate phenotypes are favoured and extreme phenotypes are selected against, the selection is said to be stabilizing. The range and distribution of phenotypes then remains approximately the same from one generation to another. Stabilizing selection is very common on the short-term scale of a few generations and when organisms remain in the same environment for a time. The individuals that survive and reproduce more successfully are those that have intermediate phenotypic values. Mortality among newborn infants, for example, is highest when they are either very small or very large; infants of intermediate size have a greater chance of surviving.

But the distribution of phenotypes in a population sometimes changes systematically in a particular direction, the so-called directional selection. The physical and biological aspects of the environment are continuously changing, and over long periods of time the changes may be substantial. The climate and even the configuration of the land or waters varies incessantly. Changes also take place in the biotic conditions; that is, in the other organisms present, whether predators, prey, parasites or competitors. Genetic changes occur as a consequence, because the fitness of different genotypes may shift so that different sets of variants are favoured. The opportunity for directional selection also arises when organisms colonize new environments, where the conditions are different from those of their original habitat. The process of directional selection often takes place in spurts. The replacement of one genetic constitution for another changes the genotypic fitnesses of genes for other traits, which in turn stimulates additional changes, and so on in a cascade of consequences.

The nearly universal success of artificial selection and the rapid response of natural populations to new environmental challenges is evidence that existing variation provides the necessary materials for directional selection, as Darwin explained. Human actions have been an important stimulus to this type of selection. Mankind transforms the environments of many organisms, which rapidly respond to the new environmental challenges through directional selection. Well-known instances are the many cases of insect resistance to pesticides, synthetic substances not present in the natural environment. As time passes, the amount of pesticide required to achieve a certain level of control must be increased again and again until finally it becomes ineffective or economically impractical. This occurs because organisms become resistant to the pesticide through directional selection. The resistance of the house fly, *Musca domestica*, to DDT was first reported in 1947. Resistance to one or more pesticides has now been recorded in more than 100 species of insects.

Sustained directional selection leads to major changes in morphology and ways of life over geologic time. Evolutionary changes that persist in a more or less continuous fashion over long periods of time are known as evolutionary trends. Directional evolutionary changes increased the cranial capacity of the human

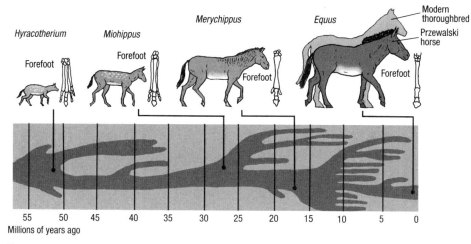

EVOLUTION OF THE HORSE, FROM *HYRACOTHERIUM*, WHICH LIVED ABOUT 55 MILLION YEARS AGO, TO THE MODERN HORSE. BRANCHES ON THE BOTTOM OF THE FIGURE REPRESENT SPECIES THAT LIVED AT DIFFERENT TIMES BUT BECAME EXTINCT.

lineage from *Australopithecus afarensis*, human ancestors who lived four million years ago, with a small brain of about 300 grams (less than a pound), to *Homo sapiens*, modern humans with a brain four times as large. The more-or-less gradual increase in size during the evolution of the horse family, from 50 million years ago to modern times, is another of the many well-studied examples of directional selection.

Two or more divergent traits in an environment may be favoured simultaneously, which is called diversifying selection. No natural environment is homogeneous; rather, the environment of any plant or animal population is a mosaic consisting of more or less dissimilar sub-environments. There is heterogeneity with respect to climate, food resources and living space. The heterogeneity may be temporal, with change occurring over time, as well as spatial, with dissimilarity found in different areas. Species cope with environmental heterogeneity in diverse ways. One strategy is the selection of a generalist genotype that is well adapted to all of the sub-environments encountered by the species. Another strategy is genetic polymorphism, the selection of a diversified gene pool that yields different genetic make-ups, each adapted to a specific sub-environment.

WHAT IS A SPECIES?
Keeping separate

$\int$pecies come about as the result of gradual change prompted by natural selection. The environments where organisms live are continuously changing over time, and they differ from place to place. Natural selection therefore favours different characteristics in different situations. The accumulation of differences eventually yields separate species.

In everyday experience we identify different kinds of organisms by their appearance. Everyone knows that people belong to the human species and are different from cats and dogs, which in turn are different from each other. There are differences among people, as well as among cats and dogs; but individuals of the same species are considerably more similar among themselves than they are to individuals of other species.

Species defined

External similarity is the common basis for identifying individuals as members of the same species, but there is more to a species than appearance. A dalmatian, a greyhound and a terrier look very different, but they are all dogs because they can interbreed. People can also interbreed with one another, and so can cats with other cats, but people cannot interbreed with dogs or cats, nor can these with each other. There is something basic, of great biological significance, behind similarity of appearance; namely, that individuals of a species are able to interbreed with one another but not with members of other species. This is expressed in the following definition: species are groups of interbreeding natural populations that are reproductively isolated from other such groups. This definition particularly applies to organisms that reproduce sexually. Sexually reproducing organisms include a majority of plants, animals and fungi. Bacteria and archaea (a major group of microscopic organisms) do not reproduce sexually, but by fission. Organisms that lack sexual reproduction are classified into different species according to criteria such as external characteristics and morphology, chemical and physiological properties and genetic make-up.

> *Species are groups of interbreeding natural populations that are reproductively isolated from other such groups.*

The ability to interbreed is of great evolutionary importance, because it determines that species are independent evolutionary units. Genetic changes originate in single individuals; they can spread by natural selection to all members of the species but not to individuals of other species. Thus, individuals of a species share a common 'gene pool' that is not shared by individuals of other species. Different species have independently evolving gene pools because they are reproductively isolated.

Chronospecies

There is no way to test whether or not today's humans could interbreed with those who lived thousands of years ago. It seems reasonable to assume that living people, or living cats, would be able to interbreed with people, or cats, similar to those that lived a few generations earlier. But what about the ancestors removed by a thousand or a million generations? The ancestors of modern humans that lived one million years ago (about 50,000 generations) are classified in the species *Homo erectus*, whereas present-day humans are classified in a different species, *Homo sapiens*; those ancestors were quite different from us in appearance, and thus it seems reasonable to conclude that interbreeding could not have occurred with modern humans.

There is not an exact time at which *Homo erectus* became *Homo sapiens*. The issue is the same as determining when an adolescent becomes an adult, or when day becomes night. It would not be appropriate to classify remote human ancestors and modern humans in the same species just because the changes from one generation to the next are small. It is useful to distinguish between the two groups by means of different species names, just as it is useful to give different names to childhood, adolescence and adulthood. Biologists distinguish species of organisms that lived at different times by means of a common-sense morphological criterion. If two organisms differ from each other about as much as two living individuals belonging to two different species, they are classified in separate species and given different names. Species that lived at different times but are related by descent are often called chronospecies (from the Greek *chronos*, time, as in chronological).

> *If two organisms differ from each other about as much as two living individuals belonging to two different species, they are classified in separate species and given different names.*

The process by which one species splits into two or more is called cladogenesis, or simply species diversification. In the archipelago of Hawaii there are about 500 species of the fruit fly *Drosophila*, which have gradually evolved over the last 20 million years; they are descendants of an ancestral species that came from North America. The cichlid fish in East African freshwater lakes is an impressive example of extreme cladogenesis, or 'adaptive radiation'. Each of the large lakes of Victoria, Malawi and Tanganyika has several hundred species that have evolved in a relatively short time from one (or perhaps two, in the case of Lake Victoria) colonizing species. Lake Victoria has about 500 cichlid species, which have evolved in less than one million years. Lake Malawi has more than 600 species, evolved within one or two million years. Lake Tanganyika, the oldest of the three, has about 200 species evolved within the last 10 million years.

Reproductive isolation

In sexually reproducing organisms, new species arise by evolution when diversification occurs in the progeny of a given species, so that eventually two (or more) reproductively isolated populations come about. Since the existence of different species is determined by their reproductive isolation, it is convenient to discuss the characteristics that account for this; they are known as reproductive isolating mechanisms (RIMs). Oaks on different islands, minnows in different rivers or squirrels in different mountain ranges cannot interbreed because they are physically separated, but not necessarily because they are biologically incompatible. Geographic separation, therefore, is not an RIM, since it is not a biological property of organisms.

There are two general categories of RIMs: prezygotic, or those that take effect before fertilization, and postzygotic, those that take effect afterward. Prezygotic RIMs prevent the formation of hybrids between members of different populations through ecological, temporal, ethological (behavioural), mechanical and gametic isolation. Postzygotic RIMs reduce the viability or fertility of hybrids and their progeny. Species that diverged quite some time ago are typically kept from interbreeding by several RIMs. As the descendants of one species become gradually divergent and eventually evolve into different species, several RIMs often come into play and accumulate in mutual reinforcement.

Origin of species

As we have established, in sexually reproducing organisms, a species is a population or a group of populations consisting of individuals able to interbreed but reproductively isolated from other populations. To ask about the origin of species is the same as asking how organisms that were able to interbreed become reproductively isolated.

Geographic speciation. In this process, two populations (or groups of populations, a specification implicit throughout the ensuing discussion) are geographically separated and evolve separately for many generations so that they become sufficiently different as to be unable to interbreed (even if they were to regain geographic proximity or overlap). In geographic isolation, natural selection does not play a role promoting reproductive isolation as such, although natural selection will occur in the separated populations as they evolve in response to different environmental conditions, including the physical components of the environment, as well as interactions with other organisms.

Geographic speciation is also known as allopatric ('different territories') speciation. The process starts as a result of geographic separation between populations. This may occur when a few colonizers reach a physically separate habitat, perhaps an island, lake, river, isolated valley or mountain range. Alternatively, a population may be split into two geographically separate ones by topographic changes, due to volcanic eruptions, earthquakes or disappearance of a water connection between two lakes, or by invasion of competitors, parasites or predators in an intermediate zone. If the process of geographic separation persists long enough, the separate populations would become genetically differentiated and eventually unable to interbreed due to the presence of one or, usually, several RIMs.

Adaptive speciation. This process assumes that gene flow between different subsets of individuals in a population may become disfavoured, as a consequence of preferences for different foods, behaviours and so on. Some initial genetic differentiation may arise due to natural selection or accidentally. If intercrossing between the two sets of individuals results in hybrids with reduced fitness, natural selection will directly promote the evolution of additional RIMs. This is because natural selection will favour genes that reduce the probability of forming hybrids.

Chromosomal speciation and polyploidy. Mutations are typically described as impacting one gene, but they can impact many genes, particularly when modifying the organization or number of chromosomes. An extreme case is polyploidy, the multiplication of entire sets of chromosomes. A diploid organism carries in the nucleus of each cell two sets of chromosomes, one inherited from each parent; a polyploid organism has three or more sets of chromosomes. Many cultivated plants are polyploid – bananas are triploid, potatoes are tetraploid, bread wheat is hexaploid, some strawberries are octaploid. These cultivated polyploids do not exist in nature, at least not in any significant frequency. Some of them first appeared spontaneously; others, such as octaploid strawberries, were intentionally produced.

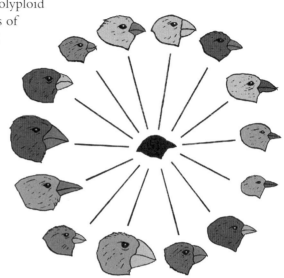

Fourteen species of Darwin's finches from the Galápagos Islands that evolved from a common ancestor. Different species have evolved beaks adapted to their different feeding habits.

In animals, polyploidy is relatively rare because it disrupts the balance between the sex chromosome and the other chromosomes, a balance required for the proper development of sex. Natural polyploid species are found in hermaphroditic animals – individuals having both male and female organs – which include snails, earthworms and planarians (a group of flatworms). They are also found in species with parthenogenetic females (which produce viable progeny without fertilization), such as some beetles, woodlice, goldfish and salamanders. Polyploid species are common among angiosperms, or flowering plants, of which about 47 per cent are polyploids. Polyploidy is rare among gymnosperms, such as pines, firs and cedars, although the redwood, *Sequoia sempervirens*, is a polyploid.

There are two kinds of polyploids: autopolyploids, which derive from a single species; and allopolyploids, which stem from a combination of chromosome sets from different species. Allopolyploid plant species are much more numerous in nature than autopolyploids.

An allopolyploid species can originate from two plant species that have the same diploid number of chromosomes. The chromosome complement of one species may be symbolized as *AA* and the other as *BB*. A hybrid of the two different species, represented as *AB*, will usually be sterile because of abnormal chromosome pairing and segregation during formation at meiosis of the gametes, which are haploid (i.e. having only half of the chromosomes, of which in a given gamete some come from the *A* set and some from the *B* set). But chromosome doubling may occur in a diploid cell as a consequence of abnormal cell division, in which the chromosomes divide but the cell does not. If this happens in the hybrid *AB*, the result is a plant cell with four sets of chromosomes, *AABB*. Such a tetraploid cell may proliferate within the plant (which is otherwise constituted of diploid cells) and produce branches and flowers of tetraploid cells. Because the flowers' cells carry two chromosomes of each kind, they can produce functional diploid gametes with the constitution *AB*. The union of two such gametes, such as happens during self-fertilization, produces a complete tetraploid individual (*AABB*). In this way, self-fertilization in plants makes possible the formation of a tetraploid individual as the result of a single abnormal cell division.

Autopolyploids originate in a similar fashion, except that the individual in which the abnormal mitosis occurs is not a hybrid. Self-fertilization thus enables a single individual to multiply and give rise to a population. This population is a new species, since polyploid individuals are reproductively isolated from their diploid ancestors. A cross between a tetraploid and a diploid yields triploid progeny, which are sterile.

WHAT ARE CHROMOSOMES, GENES AND DNA?
The double helix

*D*arwin's theory of evolution by natural selection encountered
resistance among his contemporaries and beyond because
*it was lacking an adequate theory of inheritance. In 1866 the
Augustinian monk Gregor Mendel published the fundamental
principles of the theory of heredity that is still current. Genetic
information is contained in discrete factors, or genes, which exist in
pairs, one member of the pair received from each parent.*

Mendel's theory of biological heredity became generally known
to biologists only in 1900. The next step toward understanding
the nature of genes was completed during the first quarter of the
twentieth century, when it was established that genes are parts of
the chromosomes and that they are arranged linearly along the
chromosomes. It took another quarter of a century to determine
the chemical composition of genes: DNA.

DNA
Deoxyribonucleic acid is commonly known by its abbreviated
form, DNA. Its chemical structure is a double helix made up of
two complementary strands, which are long chains of four different
kinds of nucleotides: adenine (A), cytosine (C), guanine (G), and
thymine (T). DNA has three attributes that are fundamental for life.

First, it holds the genetic information that directs all
life processes. The information is encased in sequences of the

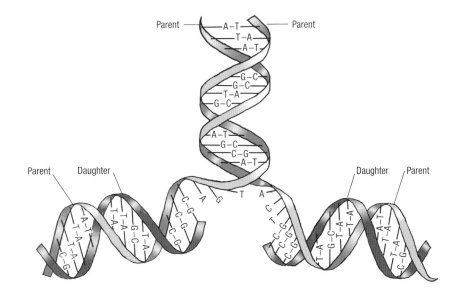

THE DNA DOUBLE HELIX. DURING REPLICATION THE TWO STRANDS UNWIND AND EACH ONE SERVES AS A TEMPLATE FOR THE SYNTHESIS OF A COMPLEMENTARY STRAND, RESULTING IN TWO DOUBLE HELICES IDENTICAL TO EACH OTHER AND TO THE ORIGINAL MOLECULE.

four nucleotides in a manner analogous to the way semantic information is conveyed by sequences of letters of the alphabet. The amount of genetic information in the DNA of an organism is enormous, because the total length of the DNA molecules is huge. For example, the human genome – that is, the DNA that each human inherits from each parent – is three billion nucleotides long. If we represent each nucleotide by one letter, as above, printing one human genome would require 1,000 books, each 1,000 pages long, with 3,000 letters (equivalent to about 500 words) per page. Scientists do not print the full genomes of humans or other organisms; rather, the DNA information is stored electronically in computers.

Second, DNA accounts for the precision of biological heredity. The two strands in the DNA double helix are complementary; both carry the same genetic information, and either one of the two strands can serve as a template for the synthesis of a complementary strand, identical to the original complementary strand. Each of the four nucleotides pairs with only one particular nucleotide in the complementary strand: A pairs only with T, and C pairs only with G. For example, if a short segment of one strand consists of the sequence ATTCAGCA, the complementary strand will be TAAGTCGT. This complementarity accounts for the fidelity of biological heredity. In the process of reproduction, the two helically coiled strands unwind, and each serves as a template for the synthesis of a complementary strand, so that the two daughter double helices are identical to each other and to the mother molecule.

The third fundamental property of DNA is mutation, which makes possible the evolution of organisms. The information encoded in the nucleotide sequence is, as a rule, faithfully reproduced during replication, so that each replication results in two DNA molecules that are identical to each other and to the maternal molecule, as explained above. The fidelity of the process is enormous but not perfect. Occasionally mutations occur in the DNA molecule during replication, so that the daughter cells differ from the parental cells (and from each other) in the nucleotide sequence or in the length of the DNA. Mutations often involve one single letter (nucleotide), but occasionally may encompass several or many letters. A mutation first appears in the DNA in a single cell of an organism, and the new, changed DNA is passed on to all cells descended from the first.

A notorious example of a mutation with important consequences in recent European history accounts for haemophilia, typically a fatal disease, which is caused by a mutation in the X chromosome. Gender in humans is determined by X chromosomes. Women have two X chromosomes; men have one X and one Y. Women with the haemophilia mutation in one X chromosome do not suffer from the disease themselves, but transmit it to those sons (half on average) who happen to inherit the mother's X chromosome with the haemophilia mutation. A mutation for haemophilia occurred in one of Queen Victoria's X chromosomes, and was transmitted through her daughters and granddaughters to the Russian, Spanish and other royal families of Europe. Tsarevich Alexis, the only son of Tsar Nicholas II of Russia, inherited haemophilia from his mother Alexandra, Queen Victoria's granddaughter. Alfonso, prince heir to the Spanish crown, inherited it from another granddaughter, Queen Ena, wife of King Alfonso XIII. Political historians believe that haemophilia contributed to the fall of the two royal families.

The mutations that count in evolution are those that occur in the sex cells because these are the cells that produce the next generation.

The mutations that count in evolution are those that occur in the sex cells (eggs and sperm), or in cells from which the sex cells derive, because these are the cells that produce the next generation. Mutations that happen in other cells will often be of little consequence and pass unnoticed. Some mutations may, however, give rise to cancer and other diseases.

Genes and chromosomes

The DNA of eukaryotic organisms (which include all animals, plants and fungi) is organized into chromosomes, which consist of several kinds of histone proteins associated with the DNA. The chromosomes occur in pairs, one inherited from each parent. The number of chromosomes characteristic of each species varies broadly, from only one pair, as in some parasitic nematodes, to more than 100, as in some species of butterflies, and more than 600, as in some ferns. Humans have 23 pairs of chromosomes. Other primates have 24 pairs; in our hominin ancestors two chromosomes fused into one, chromosome 2. In humans, as well as in primates and other mammals, the two chromosomes of a certain pair account for the individual's sex; as described above, they are identical in females (XX), but not in males (XY).

A gene is a DNA segment that becomes 'transcribed' into messenger RNA (mRNA), which in turn is 'translated' into a polypeptide; that is, a protein or part of a protein. (Some proteins consist of several polypeptides; haemoglobin A, for instance, the most common in adult humans, consists of four polypeptides, two each of two different kinds, called α and β.) RNA stands for ribonucleic acid, which is a molecule similar to DNA but usually, as in the case of mRNA, consisting of a single strand of nucleotides (rather than double-stranded, as in DNA). The RNA molecules are made up of four kinds of nucleotides. Three (A, C, G) are the same as in DNA. The fourth is uracil (U), rather than T as in DNA. The mRNA molecules are synthesized in the nucleus, as complementary to one of the DNA strands. The pairing rules between DNA and mRNA are similar to the rules of DNA, except that the A in DNA pairs with the U in mRNA (rather than with T, as in the double helix). The genetic information in the DNA is thus transported to the mRNA, which then migrates to the cytoplasm of the cell, where it determines the synthesis of proteins, following the 'genetic code', which conveys the relationship between a particular mRNA 'triplet' (sequence of three consecutive nucleotides) and a particular amino acid in the protein that is synthesized. Thus the genetic information in the DNA determines the sequence of amino acids in the protein, and consequently its molecular structure.

Enzymes may be seen as molecular machines that mediate all living processes inside cells.

There are two types of proteins. Some are important structural components of organisms; for example, collagen is the main protein component of bone. Others are enzymes, catalysts that mediate chemical reactions in organisms. Enzymes may be seen as molecular machines that mediate all living processes inside cells; that is, they catalyse the transformation of one substance into another. Enzymes are tremendously effective machines; they are thousands of millions of times more powerful than the most efficient human-made machines. The goal of so-called nanotechnology is to design molecules that may act as enzymes, and to use these enzyme-like machines to yield desired products with an efficiency immensely greater than the machines now used in human industries. Most chemical products in cells are the outcomes of a series of chain reactions, each catalysed by a different enzyme.

The number of protein-encoding genes in humans and most other animals is of the order of 20,000 to 30,000, each from several hundred to a few thousand nucleotides in length, which accounts for less than 10 per cent of the three billion DNA letters in the genome. Thus most of the DNA of eukaryotes does not embrace genes, and is therefore often called *junk* DNA. A good part of this DNA consists of sequences of various lengths, some quite small, repeated many thousands or even millions of times. Much of the junk DNA may not be functional at all, but some sequences do play a role in regulating the expression of other sequences.

The coding part of a gene often occurs in segments (exons) that are separated by sections of non-coding DNA called introns. Typically a gene is preceded by untranscribed DNA sequences, usually short, that regulate its expression. The details of how the expression of genes is controlled, how many control systems there are and how they interact remain largely to be elucidated, although great advances in evo-devo (as the evolution of developmental processes is sometimes known) have been made in recent years (see *How Do Genes Build Bodies?*).

Mendel's laws of heredity

Gregor Mendel was a monk in the Augustinian monastery in Brünn, Austria (now Brno, Czech Republic). In around 1856, he started to use garden peas (*Pisum sativum*) to investigate how individual traits are inherited. His experiments and publications are masterly examples of scientific research, even by today's standards. He published his results in 1866 in the *Proceedings of the Natural History Society* of Brünn, but they received little attention at the time from scientists anywhere.

Mendel's achievements were rediscovered in 1900 by three scientists – Hugo de Vries in Holland, Karl Correns in Germany and Erich Tschermak in Austria – who had independently obtained similar results and recognized his precedence. The significance of Mendel's principles now became apparent: the riddle of heredity was open to solution. The first step was to show that the principles also apply to animals. This was done in the early years of the twentieth century. Other important discoveries would soon follow.

Many scientists before Mendel had tried to elucidate how biological characteristics are inherited. They had crossed plants or animals and had looked at the overall similarities between the offspring and their parents. The results were confusing: the offspring resembled one parent in some traits, the other parent in others, and apparently neither one in still others. No precise regularities could be discovered.

Mendel succeeded where earlier investigators had failed, owing to his brilliant insight and methodology. He saw the need to pay attention to a single trait at a time – the shape of the seeds, or their colour, or the height of the plant, for example – rather than to the whole plant. For this purpose he selected characteristics in which the plants differed in clear-cut ways. Before starting crosses between plants, he also made sure they were true-breeding: he obtained many pea varieties from seedsmen and bred them for two years in order to select for his experiments only those strains in which the offspring always precisely resembled their parents in a given trait. Another important feature of Mendel's work was his quantitative approach: he counted the number of progeny of each kind to ascertain whether carriers of alternate traits always appeared in the same proportions.

Mendel's method of genetic analysis – counting the number of individuals of each class in the progeny of appropriate crosses – is still in use. This was the only method of genetic analysis until the discoveries of molecular biology in the 1950s. Besides his successful methodology, what made Mendel a scientific genius was his ingenuity in formulating a theory that accounted for his experimental results and in devising appropriate experimental tests to confirm it. Although properly presented as a hypothesis, Mendel's theory was formulated with an air of completeness. Time has shown that it was fundamentally correct.

In 1902 two investigators – Walter S. Sutton in the United States and Theodor Boveri in Germany – independently suggested that the Mendelian laws of inheritance could be applied to chromosomes, an idea that came to be known as the chromosome theory of heredity. Their argument was based on the parallel behaviour between chromosomes on the one hand and genes on the other. The existence of two genes for a given character, one inherited from each parent, parallels the existence of two chromosomes, also derived one from each parent. The two genes for a character segregate in the formation of the gametes because the two chromosomes of each pair pass into different gametes. Some genes for different characters assort independently because they are in non-homologous (different) chromosomes, and these chromosomes group themselves in the gametes independently of the parent from which they came.

The double helix

DNA and RNA are both nucleic acids. The nucleic acids were first described by F. Miescher in 1874. In 1944, Oswald Avery, Colin MacLeod and Maclyn McCarty obtained results suggesting that DNA was the carrier of hereditary information. They obtained highly purified DNA from a virulent strain of pneumococcus (*Diplococcus pneumoniae*). Non-virulent pneumococci were incubated in the presence of this purified DNA, and some virulent pneumococci were recovered. A genetic transformation had occurred; DNA was described as a transforming agent. In 1949, A.E. Mirsky and Hans Ris found that all somatic cells of an organism contain, as a rule, the same amount of DNA, while gametic cells contain half as much as somatic cells. This is what would be expected of the genetic material. Proteins were at that time favoured by many biologists as the substances most likely to encode hereditary information. Mirsky and Ris found different amounts of them in different cells of a given organism.

Although properly presented as a hypothesis, Mendel's theory was formulated with an air of completeness. Time has shown that it was fundamentally correct.

Direct evidence of DNA as hereditary material was obtained in 1952 by Alfred D. Hershey and Martha Chase. They demonstrated that when the bacteriophage virus *T2* infects the bacterium *Escherichia coli*, only the DNA of the phage enters the bacteria and brings about its own replication. The protein coat that envelops the virus DNA neither enters the bacteria nor participates in the replication process.

James Watson and Francis Crick proposed in 1953 the double-helix model of DNA. This model was consistent with all the information then available about the composition and structure of DNA, and provided a plausible explanation of the basic properties of the hereditary material – carrier of information and binary replication. The double-helix model of DNA has since been confirmed.

The hereditary material of all organisms, except some viruses, is DNA organized in a double helix. Several kinds of viruses are known to contain single-stranded DNA. Most RNA-containing viruses have single-stranded RNA, although some have duplex RNA molecules.

Watson and Crick pointed out that the double helix suggests a mechanism for the precise replication of genes. The sequence along one of the strands unambiguously specifies the sequence along the complementary strand, owing to the strict determination of the base-pairing between the two DNA chains (A with T and C with G). If the two strands in a given molecule were to separate by an unwinding process and become exposed to a pool of nucleotides, each might serve as a template for a complementary strand. Two new double helices would be formed, which, because of the rules of base-pairing, would be identical to each other and to the parental double helix.

HOW DO GENES BUILD BODIES?
Egg to adult, brain to mind

O *ntogenetic decoding, or the egg-to-adult transformation and the brain-to-mind puzzle, are two great research frontiers facing biology in the twenty-first century. Ontogenetic decoding refers to the problem of how the unidimensional genetic information encoded in the DNA of a single cell becomes transformed into a four-dimensional being: the individual that grows, matures and dies. The brain-to-mind puzzle encompasses two interdependent issues: first, how the physicochemical signals that reach our sense organs become transformed into perceptions, feelings, ideas, critical arguments, aesthetic emotions and ethical values; second, how out of this diversity of experiences there emerges a unitary reality, the mind or self.*

The benefits that the elucidation of ontogenetic decoding would bring to mankind are enormous. It would enable us to comprehend the modes of action of complex genetic diseases, including cancer, and therefore their cure. It would also bring an understanding of the process of ageing, the unforgiving disease that kills all those who have won the battle against other infirmities.

Cancer is an anomaly of ontogenetic decoding: cells proliferate even though the welfare of the organism demands otherwise. Individual genes (oncogenes) have been identified as being involved in the causation of particular forms of cancer. Ageing is also a failure of the process of ontogenetic decoding:

cells fail to carry out the functions imprinted in their genetic codescript or are no longer able to proliferate and replace dead cells.

Mendel to Dolly

The instructions that guide the ontogenetic process, or the egg-to-adult transformation, are carried in the hereditary material. The first important step towards understanding how the genetic information is decoded came in 1941, when George W. Beadle and Edward L. Tatum demonstrated that genes determine the synthesis of enzymes; enzymes are the catalysts that control all chemical reactions in living beings. Later it became known that amino acids (the components that make up enzymes and other proteins) are encoded by a set of three consecutive nucleotides. This relationship accounts for the linear correspondence between a particular sequence of coding nucleotides and the sequence of the amino acids that make up the encoded enzyme.

The instructions that guide the ontogenetic process, or the egg-to-adult transformation, are carried in the hereditary material.

Chemical reactions in organisms must occur in an orderly manner; organisms must have ways of switching genes on and off since different sets of genes are active in different cells. The first control system was discovered in 1961 by François Jacob and Jacques Monod for a gene encoding an enzyme that digests sugar in the bacterium *Escherichia coli*. The gene is turned on and off by a system of several switches consisting of short DNA sequences adjacent to the coding section of the gene. (The coding sequence of a gene is the part that determines the sequence of amino acids in the encoded enzyme.) The switches acting on a given gene are activated or deactivated by feedback loops that involve molecules synthesized by other genes. A variety of gene control mechanisms were soon discovered, in bacteria and other microorganisms. Two elements are typically present: feedback loops and short DNA sequences acting as switches. The feedback loops ensure that the presence of a substance in the cell induces the synthesis of the enzyme required to digest it, and that an excess of the enzyme in the cell represses its own synthesis. (For example, the gene encoding a sugar-digesting enzyme in *E. coli* is turned on or off by the presence or absence in the cell of the sugar to be digested.)

The investigation of gene control mechanisms in insects, mammals and other complex organisms became possible in the mid 1970s with the development of recombinant DNA techniques. This technology made it feasible to isolate single genes (and other DNA sequences) and to multiply, or 'clone', them billions of times over, in order to obtain the quantities necessary for ascertaining their nucleotide sequence and other properties. One unanticipated discovery was that most genes come in pieces: the coding sequence of a gene is divided into several fragments, separated one from the next by non-coding DNA segments. In addition to the alternating succession of coding (exons) and non-coding (introns) segments, mammalian genes contain short control sequences, like those in bacteria but typically more numerous and complex, that act as control switches and signal where the coding sequence begins.

Much remains to be discovered about the control mechanisms of mammalian genes. The daunting speed at which molecular biology is advancing has led to the discovery of some prototypes of insect and mammalian gene control systems, but there is still a great deal to be unravelled. Moreover, understanding the control mechanisms of individual genes is only the first step towards solving the mystery of ontogenetic decoding. The second is the puzzle of differentiation.

Egg to adult

A human being consists of one trillion cells of some 300 different kinds, all derived by sequential division, each cell dividing into two, from the fertilized egg, a single cell 0.1 millimetres in diameter. The first few divisions yield a spherical mass of amorphous cells. Successive divisions are accompanied by the appearance of folds and ridges in the mass of cells, and later on by the variety of tissues, organs and limbs characteristic of a human individual. The full complement of genes duplicates with each cell division, so that two complete genomes are present in every cell. Yet different sets of genes are active in different cells. This must be so in order for cells to differentiate: a nerve cell, a muscle cell and a skin cell are vastly different in size, configuration and function. Nevertheless, experiments with other animals (and some with humans) indicate that all the genes in any cell have the potential to become activated. (The sheep Dolly was conceived using the genes extracted from a cell in an adult sheep.)

The information that controls cell and organ differentiation is ultimately contained in the DNA sequence, but mostly in short segments and a limited number of genes. In mammals, insects and other complex organisms, there are control circuits operating at higher levels than the mechanisms that activate and deactivate individual genes. These higher-level circuits (such as the so-called homeobox genes) act on sets rather than individual genes. The details of how these sets are controlled, how many control systems there are and how they interact are some of the many questions that need to be resolved in order to elucidate the egg-to-adult transformation. The advances in knowledge over the last two decades have been impressive. Experiments with stem cells are likely to provide important additional knowledge as scientists ascertain how stem cells become brain cells in one case but muscle cells in another, and how some cells become the heart and others the liver.

The genetic toolkit

A successful way of discovering the details of how genes determine the formation and patterning of an animal has been the isolation of single-gene mutations that distort the animal's body plan. Early studies carried out on *Drosophila* flies and other insects discovered individual genes and sets of genes that impacted specific morphological features. It was soon discovered that the genetic toolkit that determines the patterning of the body plan and parts in vertebrates and other animals was similar to the genetic toolkit for insect development.

In animal genomes, most genes are 'housekeeping' genes that encode the enzymes and other proteins engaged in the essential processes that occur in cells, whether in all cells, as with the genes that regulate metabolism, or in specific cells, such as those in the immune system or the blood cells. The genetic toolkit that determines the body plan and the patterning of organs consists of different sorts of genes. There are two main sorts of toolkit genes: some encode transcription factors that regulate the expression (turning on and off) of many other genes during development; others are engaged in signalling pathways that mediate interactions between cells.

One set of toolkit genes encoding transcription factors are the *Hox* (homeobox) genes, which in *Drosophila* are a set of eight genes arranged along the chromosome in a sequence that linearly correspond to the sequence of the body parts that they impact. That is, genes engaged in pattern formation in the head are followed by genes impacting the thorax, which in turn are followed by genes that impact the abdomen and caudal morphology. These genes were first identified in *Drosophila* through mutations that resulted in gross abnormal morphologies, such as antennae with the configuration of legs, or flies with a double thorax and four wings, rather than the normal single thorax and two wings.

The *Hox* genes in vertebrates and other animal phyla are somewhat similar and likewise arranged in sequence, but there are more genes arranged in several sets. In the mouse, for example, there are 39 *Hox* genes, distributed in four different chromosomes, with the genes in each chromosome linearly arranged in sequences that largely parallel the sequence and functions of the *Drosophila Hox* genes.

The *Drosophila* genes engaged in metabolic signalling pathways also have similar corresponding genes in the mouse, human and other vertebrates. Because pathway-signalling genes were first discovered in *Drosophila* through mutants with abnormal morphologies, they are often exotically named by reference to the resulting abnormality, such as *wingless, Delta, Serrate, Notch, hedgehog* and *cubitus interruptus*. Humans, vertebrates and many other animals have 20,000 genes. The total number of toolkit genes is estimated to be about 200 in the nematode *Caenorhabditis elegans*, about 400 in *Drosophila* and nearly 900 in humans. In all three organisms a large majority of toolkit genes encode transcription factors, while 10 per cent or less are engaged in signalling pathways.

The overall conservation of the genetic toolkit across animal phyla is remarkable. It surely reflects a common evolutionary ancestry. But questions remain yet to be solved. The eye of an insect and the eye of a vertebrate are very different. An insect, for example *Drosophila*, has a compound eye made up of several hundred ommatidia, consisting each of one central cell and eight surrounding photoreceptor cells. Humans and other vertebrates have a 'camera eye', with a complex organization of numerous parts, including the cornea, lens, retina, optic nerve, muscle and much more. Yet the human *Pax6* gene and the *Drosophila eyeless* are orthologous genes (i.e. they derive from the same ancestral gene) acting at the top of the regulatory process that directs eye development in two entirely different animals, which also share other genes engaged in eye development, such as *sine oculis*, *eyes absent* and *opsin*.

Brain to mind

The brain is the most complex and most distinctive human organ. It consists of 30 billion nerve cells, or neurons, each connected to many others through two kinds of cell extensions, known as the axon and the dendrites. From the evolutionary point of view, the animal brain is a powerful biological adaptation; it allows the organism to obtain and process information about environmental conditions and then to adapt to them. This ability has been carried to the limit in humans; the extravagant hypertrophy of the human brain makes possible free will and language, social and political institutions, technology and art.

The most rudimentary ability to gather and process information about the environment is found in certain single-celled microorganisms. The protozoan *Paramecium* swims apparently at random, ingesting the bacteria it encounters, but when it meets unsuitable acidity or salinity, it checks its advance and starts in a new direction. The single-celled alga *Euglena* not only avoids unsuitable environments but seeks suitable ones by orienting itself according to the direction of light, which it perceives through a light-sensitive spot in the cell. Plants have not progressed much further. Except for those with tendrils that twist around any solid object and the few carnivorous plants that respond to touch, they mostly react only to gradients of light, gravity and moisture.

In animals the ability to secure and process environmental information is mediated by the nervous system. The simplest nervous systems are found in corals and jellyfishes; they lack coordination between different parts of their bodies, so that any one part is able to react only when it is directly stimulated. Sea urchins and starfish possess a nerve ring and radial nerve cords that coordinate stimuli coming from different parts; hence they respond with direct and unified actions of the whole body. They have no brain, however, and seem unable to learn from experience. Planarian flatworms have the most rudimentary brain known; their central nervous system and brain process and coordinate information gathered by sensory cells. These animals are capable of simple learning and hence of variable responses to repeatedly encountered stimuli. Insects and their relatives have much more advanced brains; they obtain precise chemical, acoustic, visual and tactile signals from the environment and process them, making possible complex behaviours, particularly in searching for food and selecting mates.

The animal brain is a powerful biological adaptation; it allows the organism to obtain and process information about environmental conditions and then to adapt to them.

Vertebrates – animals with backbones – are able to obtain and process much more complicated signals and to respond to the environment more variably than insects or any other invertebrates. The vertebrate brain contains an enormous number of associative neurons arranged in complex patterns. In vertebrates, the ability to react to environmental information is correlated with an increase in the relative size of the cerebral hemispheres and of the neopallium, an organ involved in associating and coordinating signals from all receptors and brain centres. In mammals, the neopallium has expanded and become the cerebral cortex. Humans have a very big brain relative to their body size, and a cerebral cortex that is disproportionately large and complex even for their brain size. Abstract thinking, symbolic language, complex social organization, values and ethics are manifestations of the wondrous capacity of the human brain to gather information about the external world and to integrate that information and react flexibly to what is perceived.

Within the last two decades, neurobiology has developed into one of the most exciting biological disciplines. An increased commitment of financial and human resources has brought an unprecedented rate of discovery. Much has been learned about how light, sound, temperature, resistance and chemical impressions received in our sense organs trigger the release of chemical transmitters and electric potential differences that carry the signals through the nerves to the brain and elsewhere in the body. A great deal has also been learned about how neural channels for information transmission become reinforced by use or may be replaced after damage; about which neurons or groups of neurons are committed to processing information derived from a particular organ or environmental location; and about many other matters. But for all this progress, neurobiology remains an infant discipline, at a stage of theoretical development comparable perhaps to that of genetics at the beginning of the twentieth century. Those things that count most remain shrouded in mystery: how physical phenomena become mental experiences, and how out of the diversity of these experiences emerges the mind, a reality with unitary properties, such as free will and the awareness of self, that persist through an individual's life.

These mysteries are not unfathomable; rather, they are puzzles that the human mind can tackle with the methods of science. Over the next half-century or so, many of these puzzles, like those that modulate the egg-to-adult transformation, will be solved. We will then be well on our way towards answering the injunction: 'Know thyself.'

WHAT IS MOLECULAR EVOLUTION?
The molecular clock

*M*olecular evolution is the biological discipline that studies evolution using the methods of molecular biology. It emerged in the mid twentieth century, following the 1953 discovery of the double-helix structure of DNA, the hereditary chemical. Molecular biology provides the strongest evidence of biological evolution and makes it possible to reconstruct the evolutionary history of living species with as much detail and precision as anyone might want.

In all organisms from bacteria and protozoa to plants and animals, the chemical components of life are the same, and occur in very similar proportions. This uniformity makes sense if it is due to a common origin. In reconstructing evolutionary history, it is all-important that genetic information is stored in the linear array of letters (nucleotides) that make up the DNA. The DNA sequences from different organisms can then be aligned, and the number of letters that are different between them reflect the time elapsed since their last common ancestor.

The unity of life
Molecular biology, a discipline that emerged nearly 100 years after the publication of *On the Origin of Species*, has provided the strongest evidence yet for the evolution of organisms. It proves evolution in two ways: first by showing the unity of life in the nature of DNA and the workings of organisms at the level of enzymes and other protein molecules; second, and most important

for evolutionists, by making it possible to reconstruct evolutionary relationships that were previously unknown, and to confirm, refine and time evolutionary relationships from the universal common ancestor up to all living organisms. The precision with which these events can be reconstructed is one reason why the evidence from molecular biology is so useful to evolutionists and so compelling.

The molecular components of organisms, how they are assembled and the metabolic pathways of cells are remarkably similar (see *How Did Life Begin?*). The genetic code by which the information contained in the DNA of the cell nucleus is passed on to proteins is the same virtually everywhere. Similar metabolic pathways – sequences of biochemical reactions – are used by the most diverse organisms to produce energy and to make up the cell components. Many other pathways are theoretically possible, but only a limited number are used, and the pathways are the same in organisms with extremely different ways of life.

In all organisms from bacteria and protozoa to plants and animals, the chemical components of life are the same, and occur in very similar proportions.

Molecular information

Genes and proteins are long molecules that contain information in the sequence of their components in the same way that the meaning of sentences is conveyed by the sequence of letters and words. The sequences that make up the genes are passed on from parents to offspring and are identical from generation to generation, except for occasional changes introduced by mutations. Closely related species have very similar DNA sequences; the few differences reflect mutations that have occurred since their last common ancestor. Species that are less closely related to one another exhibit more differences in their DNA because more time has elapsed since their last common ancestor. This is the rationale used for reconstructing evolutionary history using molecules.

Genes and proteins are long molecules that contain information in the sequence of their components.

Molecular biology offers two kinds of arguments for evolution. Using the alphabet analogy, the first argument says that languages that use the same alphabet (the same hereditary molecule, the DNA made up of the same four nucleotides and the same 20 amino acids in their proteins) as well as the same dictionary (the same genetic code) cannot be of independent origin. The second argument concerns the degree of similarity in the sequence of nucleotides in the DNA (and thus the sequence of amino acids in the proteins). The degree of similarity between sequences of DNA (and protein) is what makes it possible to reconstruct evolutionary history using molecular information.

Informational macromolecules

DNA and proteins are called 'informational macromolecules' because they are long linear molecules made up of sequences of units – nucleotides in nucleic acids, amino acids in proteins – that embody evolutionary information. Comparing the sequence of the components in two macromolecules establishes how many units are different. Because evolution usually occurs by changing one unit at a time, the number of differences is an indication of the recency of common ancestry. Thus, the inferences from palaeontology, comparative anatomy and other disciplines that study evolutionary history can be tested in molecular studies of DNA and proteins by examining the sequences of nucleotides and amino acids. The authority of this kind of test is overwhelming: each of the thousands of genes and thousands of proteins contained in an organism provides an independent test of that organism's evolutionary history.

The degree of similarity in a sequence of nucleotides or amino acids can be precisely quantified. For example, in humans and chimpanzees, the protein molecule called cytochrome *c*, which serves a vital function in respiration within cells, consists of the same 104 amino acids in exactly the same order. It differs, however, from the cytochrome *c* of rhesus monkeys by one amino acid, from that of horses by 11 amino acids, and from that of tuna by 21 amino acids.

Molecular evolutionary studies have three notable advantages over comparative anatomy and the other classical disciplines: precision, universality and multiplicity. Precision exists because the information is readily quantifiable. The number of units that are different is easily established when the sequence of units is known for a given macromolecule in different organisms. It is simply a matter of aligning the units (nucleotides or amino acids) between two or more species and counting the differences.

The second advantage is universality: comparisons can be made between very different sorts of organisms. There is little that comparative anatomy can say when, for example, organisms as diverse as yeasts, pine trees and human beings are compared, but there are numerous DNA and protein sequences that can be compared in all three.

The third advantage is multiplicity. Each organism possesses thousands of genes and proteins, every one of which reflects the same evolutionary history. If the investigation of one particular gene or protein does not satisfactorily resolve the evolutionary relationship of a set of species, additional genes and proteins can be investigated until the matter has been settled.

Moreover, the widely varying rates of evolution among sets of genes opens up the opportunity for investigating different genes in order to achieve different degrees of resolution in the tree of evolution. Evolutionists rely on slowly evolving genes for reconstructing remote evolutionary events, but increasingly faster-evolving genes for reconstructing the evolutionary history of more recently diverged organisms.

Genes that encode ribosomal RNA molecules are among the slowest-evolving genes. (Ribosomes are complex molecules that mediate the synthesis of proteins; each ribosome consists of many proteins and several RNA molecules.) They have been used to reconstruct the evolutionary relationships among groups of organisms that diverged a long time ago: for example, bacteria, archaea and eukaryotes (the three major divisions of the living world), which diverged more than two billion years ago (see page 105), or the microscopic protozoa (e.g. Plasmodium, which causes malaria) compared with plants and animals, groups of organisms that diverged about a billion years ago. Cytochrome c, mentioned earlier, evolves slowly, but not as slowly as the ribosomal RNA genes. Thus it is used to decipher the relationships within large groups of organisms, such as humans, fishes and insects. Fast-evolving molecules, such as the fibrinopeptides involved in blood clotting, are appropriate for investigating the evolution of closely related animals – the primates, for example: macaques, chimps and humans.

Molecular evolutionary trees

DNA and proteins provide information not only about the branching of lineages from common ancestors (diversification, or cladogenesis), but also about the amount of genetic change that has occurred in any given lineage (anagenesis). It might seem at first that quantifying the number of changes that have occurred in the past evolution of a lineage would be impossible because it would require comparison of molecules from organisms that are now extinct with molecules from living organisms; for example, we might want to know how much cytochrome c may have changed from their common ancestor to humans and monkeys. Organisms of the past are sometimes preserved as fossils, but their DNA and proteins have largely disintegrated. However, comparisons between living species can provide information about change through time.

DNA and proteins provide information not only about the branching of lineages from common ancestors, but also about the amount of genetic change that has occurred in any given lineage.

Consider again, for example, the protein cytochrome c, involved in cell respiration. In animals, cytochrome c consists of 104 amino acids. When the amino acid sequences of humans and rhesus monkeys are compared, they are found to be different at position 58 – isoleucine (I) in humans, threonine (T) in rhesus monkeys – but identical at the other 103 positions (see the figure opposite, where dots indicate amino acids identical to those of human cytochrome c).

```
Human            G - D - V - E - K - G - K - K - I - M - K - C - S - Q - C -
Rhesus monkey    •   •   •   •   •   •   •   •   •   •   •   •   •   •   •
Horse            •   •   •   •   •   •   •   •   V   Q   •   •   A   •   •
```

```
H - Y - V - E - K - G - G - K - H - K - Y - G - P - N - L - H - G - L - F - G - R - K - T -
•   •   •   •   •   •   •   •   •   •   •   •   •   •   •   •   •   •   •   •   •   •   •
•   •   •   •   •   •   •   •   •   •   •   •   •   •   •   •   •   •   •   •   •   •   •
```

```
G - Q - A - P - G - Y - S - Y - T - A - A - N - K - N - K - N - K - G - I - W - G - E - D -
•   •   •   •   •   •   •   •   •   •   •   •   •   •   •   •   •   T   •   •   •   •
•   •   •   •   •   F   T   •   D   •   •   •   •   •   •   •   •   T   •   K   •   E
```

```
K - L - M - E - Y - L - E - N - P - K - K - Y - I - P - G - T - K - M - I - F - V - G - I - K -
•   •   •   •   •   •   •   •   •   •   •   •   •   •   •   •   •   •   •   •   •   •   •
•   •   •   •   •   •   •   •   •   •   •   •   •   •   •   •   •   •   •   •   A   •   •
```

```
K - K - E - E - R - A - D - L - I - A - Y - L - K - K - A - Y - N - E
•   •   •   •   •   •   •   •   •   •   •   •   •   •   •   •   •
•   •   T   •   •   E   •   •   •   •   •   •   •   •   •   •   •
```

When humans are compared with horses, 12 amino acid differences are found, and when horses are compared with rhesus monkeys, there are 11 amino acid differences. Even without knowing anything else about the evolutionary history of mammals, we would conclude that the lineages of humans and rhesus monkeys diverged from each other much more recently than they diverged from the horse lineage. Moreover, it can also be concluded that the amino acid difference between humans and rhesus monkeys must have occurred in the human lineage after its separation from the rhesus monkey lineage. This conclusion is reached because monkey and horse (as well as other animals) have the same amino acid (T) at this position, while the human is different (amino acid I).

The amino acid sequences in the cytochrome *c* of 20 very diverse organisms were ascertained in 1967. Counting the amino acid differences between the species (and using appropriate statistical methods to determine the likely relations among species and sets of species) resulted in the evolutionary tree shown in the figure. This was the first important instance of molecular evolution data being used to reconstruct the evolutionary tree of organisms that diverged from one another hundreds of millions of years ago. The common ancestor (at the bottom) of yeast and humans lived more than a billion years ago.

The reconstruction of evolutionary history accomplished with DNA and protein molecules follows the same logic used in comparative anatomy and other traditional methods. In palaeontology, the time sequence of fossils is determined by the age of the rocks in which they are embedded. As pointed out above, each of the thousands of genes and thousands of proteins contained in an organism provides an independent test of that organism's evolutionary history. Many thousands of tests have been done (and thousands more are published every year); not one has given evidence contrary to evolution. There is probably no other notion in any field of science that has been as extensively tested and as thoroughly corroborated as the evolutionary origin of living organisms.

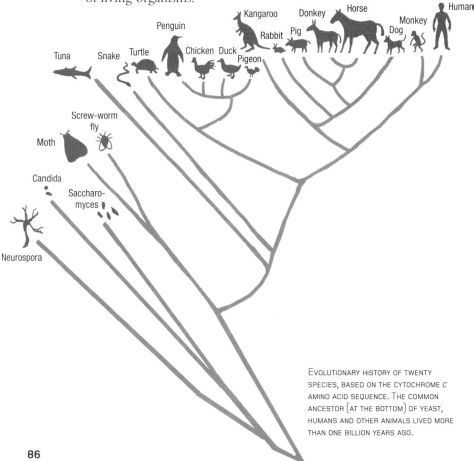

EVOLUTIONARY HISTORY OF TWENTY SPECIES, BASED ON THE CYTOCHROME *c* AMINO ACID SEQUENCE. THE COMMON ANCESTOR (AT THE BOTTOM) OF YEAST, HUMANS AND OTHER ANIMALS LIVED MORE THAN ONE BILLION YEARS AGO.

The molecular clock

One conspicuous attribute of molecular evolution pointed out above is that differences between homologous molecules can readily be quantified and expressed as, for example, proportions of nucleotides or amino acids that have changed. (Homologous molecules are those derived from a common ancestral molecule.) Rates of evolutionary change can therefore be more precisely established with respect to DNA or proteins than with respect to traits such as the configuration or function of an organ or limb. Studies of molecular evolution rates have led to the proposition that macromolecules may serve as evolutionary clocks.

It was first observed in the 1960s that the number of amino acid differences between homologous proteins of any two given species seemed to be roughly proportional to the time of their divergence from a common ancestor. If the rate of evolution of a particular protein or gene were approximately the same in the evolutionary lineages leading to different species, proteins and DNA sequences would provide a molecular clock of evolution. They could then be used to reconstruct not only the sequence of branching events of a phylogeny but also the time at which the various events occurred.

Consider, for example, cytochrome c. If the substitution of nucleotides in the gene coding for cytochrome c occurred at a constant rate through time, we could determine the time elapsed along any branch of the phylogeny simply by examining the number of nucleotide substitutions along that branch. We would need to calibrate the clock by reference to an outside source, such as the fossil record, that would provide the actual geologic time elapsed in at least one specific lineage or since one particular branching point. For example, if the time of divergence between insects and vertebrates is determined to have occurred 700 million years ago, other times of divergence can be determined by proportion of the number of amino acid changes.

The molecular evolutionary clock is not expected to be a metronomic clock, like a watch that measures time exactly, but a stochastic (probabilistic) clock, like radioactive decay. In a stochastic clock the probability of a certain amount of change is constant (e.g. a given quantity of atoms of carbon-14 is expected, through decay, to be reduced by half in 5,730 years, its 'half-life'), although some variation occurs in the actual amount of change. Over fairly long periods of time a stochastic clock can be quite accurate. The enormous potential of the molecular evolutionary clock lies in the fact that every gene and protein is a separate clock. The clocks 'tick' at a different rate – the rate of evolution characteristic of a particular gene or protein – but each of the thousands and thousands of genes or proteins provides an independent measure of the same evolutionary events.

Evolutionists have found that the amount of variation observed in the evolution of DNA and proteins is greater than is expected from a stochastic clock – in other words, the clock is 'overdispersed', or somewhat erratic. The discrepancies in evolutionary rates along different lineages are not excessively large, however. So although it is possible to time phylogenetic events with considerable accuracy, more genes or proteins (about two to four times as many) must be examined than would be required if the clock were stochastically constant, in order to achieve a desired degree of accuracy. The average rates obtained for several proteins taken together become a fairly precise clock, particularly when many species are studied.

HOW DID LIFE BEGIN?
A chicken-and-egg question

*B*iologists *agree that life on our planet originated spontaneously by natural processes from the same chemicals of which living organisms consist today, such as carbon, nitrogen, oxygen and hydrogen. Does that mean that we know how life began? Not quite. Although there are some good ideas and experiments about the origin of life, there is not as yet general agreement as to how it might have started. What we do know, because the evidence is overwhelming, is that all living organisms on Earth have evolved from a single original form of life.*

The diversity of life on Earth is staggering. More than one million existing species of plants and animals have been named and described; many more remain to be discovered, at least ten million according to most estimates. What is impressive is not just the numbers, but also the incredible heterogeneity in size, shape and way of life: from lowly bacteria less than one thousandth of a millimetre in diameter, to the stately sequoias of California, rising 100 metres above the ground and weighing several thousand tons; from microorganisms living in the hot springs of Yellowstone National Park at temperatures near the boiling point of water, some like *Pyrolobus fumarii* able to grow at more than 100°C, to fungi and algae thriving on the ice masses of Antarctica and in saline pools at -23°C; from the strange wormlike creatures discovered in dark ocean depths, thousands of metres below the surface, to spiders and larkspur plants existing on Mount Everest more than 6,600 metres above sea level.

The unity of life

The exact number of species on Earth is not known, though estimates range between 10 million and 30 million. That colossal wealth of species and their stupendous diversity is the outcome of the evolutionary processes, starting from a single life form.

The molecular components of organisms are remarkably uniform – in the nature of the components as well as the ways in which they are assembled and used. In all organisms, from bacteria and archaea to plants, animals and humans, the instructions that guide development and functioning are encased in the same hereditary material, DNA, which provides the instructions for the synthesis of proteins. The thousands of enormously diverse proteins that exist in organisms are synthesized from different combinations, in sequences of variable length, of 20 amino acids, the same in all proteins and all organisms. Yet several hundred other amino acids exist in nature. There is a very long list of features that could possibly take alternative configurations and yet are uniform throughout all life.

> *The molecular components of organisms are remarkably uniform – in the nature of the components as well as the ways in which they are assembled and used.*

Moreover, the complex machinery by which the hereditary information is conveyed from the nucleus to the main body of the cell is everywhere the same: the sequence of nucleotides in the DNA is transcribed into a complementary sequence of RNA (dubbed 'messenger RNA'), which becomes translated into specified sequences of amino acids that make up the proteins and enzymes that carry out all life processes. This translation involves specific RNA molecules ('transfer RNA') and RNA-protein complexes (ribosomes), universally shared. The genetic dictionary that guides the translation of the mRNA sequence into the amino acid sequence of proteins is also universal.

The unity of life – i.e. the similarity of components and processes in all organisms – reveals the genetic continuity and common ancestry of all organisms. Consider the genetic code as an example. Each particular sequence of three nucleotides (codon) in the nuclear DNA acts as a code for exactly the same amino acid in all organisms. For example, in any given gene of any organism, the codon GCC determines that the amino acid alanine will be incorporated in the protein specified by the gene; the codon GAC determines the incorporation of the amino acid asparagine, and so on. The universal correspondence between the DNA language (codons) and the protein language (amino acids) is analogous with two spoken languages using the same combination of letters for representing the same particular concept or object. If we find that certain sequences of letters – planet, tree, woman – are used with identical meaning in a number of different books, we can be sure that they have a common origin.

Early life

When did life begin? The oldest known rocks on Earth are 4.3 billion years old, only a few hundred million years younger than the Sun, which is thought to have formed somewhat earlier than 4.5 billion years ago. Life originated on Earth several hundred million years later, at least 3.4 billion years ago. Stromatolites are structures formed by columns of cyanobacteria, a type of photosynthetic bacteria. Ancient structures that resemble stromatolites have been discovered in Australia, Canada and Baja California, Mexico. Some of these stromatolite-like structures have been dated by radiometric methods at approximately 3.5 billion years. Electron microscopy has revealed structures that resemble cyanobacteria in some stromatolites.

The oldest fossils ever found, dated as 3.4 billion years old, were discovered recently between cemented sand grains from an ancient beach in Western Australia. Other sources of evidence from different parts of the world indicate various forms of microscopic bacteria-like organisms that are at least 3 billion years old. It is commonly accepted by experts that life originated on Earth during a 300-million-year interval between 3.8 and 3.5 billion years ago.

What is life?

Before we can answer the question 'How did life begin?', we need to address a related question, 'What is life?'. Two essential properties of life are heredity and metabolism. Cells reproduce by dividing, by making copies of themselves. To ensure continuity of life, the daughter cells must inherit the same components that make up the mother cell. These components include instructions about the chemical machinery of the cell: what chemicals to make and how they will operate. The instructions (DNA) are themselves encased in chemicals. But the synthesis of the instruction chemicals requires the chemical machinery of the cell. We have a chicken-and-egg problem to get the process started. The instructions detail how the chemical machinery will operate, but the instructions cannot be synthesized *without* the chemical machinery. Think of a document to be copied: you need the photocopy machine to do the copying. Better yet, think of electronic computers: you need the hardware (the computer) to carry out the instructions and the software to provide the instructions about how to make the computer. In life, the software includes the instructions (DNA) as to how to make the computer (the cell's machinery).

The information-carrying constituents are DNA (deoxyribonucleic acid) and RNA (ribonucleic acid) molecules. DNA and RNA are strings made up of four kinds of chemical components, represented by the letters A, C, G and T in DNA, and by A, C, G and U in RNA; that is, U replaces T in RNA. As described in *What Are Chromosomes, Genes and DNA?*, the hereditary information is contained in long sequences of the four letters, in the same way that semantic information is contained in sequences of the letters of the alphabet. The machinery that carries out the chemical reactions (called 'metabolism') consists of enzymes, which are proteins able to catalyse chemical reactions with great accuracy and at speeds much faster than any human-made machine. Proteins are made up of long strings of 20 different kinds of components, called amino acids.

> *A cell consists of thousands and thousands of components, including many thousands of different kinds of enzymes, carrying out with tremendous efficiency and precision many thousands of chemical reactions.*

A cell consists of thousands and thousands of components, including many thousands of different kinds of enzymes, carrying out with tremendous efficiency and precision many thousands of chemical reactions, a great number of them precisely determined sequences of reactions. Graphic representations of cellular processes show them as extremely complex networks, resembling something like a map of the transportation system in the United States. There are interstate and state highways, plus a grid of additional roads, trails, driveways and access roads; oil wells, refineries, gas stations and factories; cars, trucks, motorcycles and other vehicles; rivers, waterways, harbours and all sorts of boats; airports and aeroplanes.

Consider, now, the question: how did the transportation network of the United States get started? We might think of simple foot trails, followed later by roads, paved or otherwise, suitable for carts and wagons. But it would be difficult to ascertain where the first trails were fashioned, and which towns or villages they connected. The first trails in North America were made a few thousand years ago. In contrast, life originated on Earth a few thousand *million* years ago. So determining the origins of the trails of life is all the more difficult.

Life in the laboratory?

If we want to know how life first started, we need to identify the primitive constituents that made up the simplest forms of life. From the start we encounter, as pointed out, the chicken-and-egg problem. We need information molecules, such as DNA or RNA, which convey from mother cell to daughter cells the information as to what enzymes are to be synthesized and how to synthesize them. But the DNA or RNA molecules need to be synthesized themselves, and for that we need metabolism, the chemical machinery of living processes. Returning to the computer metaphor, in life the software has the information as to how the computer is to be built, but the software (heredity molecules) cannot be read without the computer (metabolic machinery). Once you have computers in existence, there is no problem. The problem is how to get the first computer.

In 1953, Stanley Miller, a graduate student in chemistry at the University of Chicago, simulated in a tabletop glass apparatus the conditions that might have occurred on our planet soon after its birth. He included some inorganic chemicals that may have been present in the early Earth, such as ammonia, methane and hydrogen, and added water vapour and electric discharges to simulate lightning, as that may also have occurred at that time. After one week, amino acids, including many found in modern proteins, as well as other compounds, such as urea, naturally occurring only in organisms, had formed in the five-litre glass flask in which the experiment was conducted. Miller had thus demonstrated that organic compounds could be formed without the mediation of enzymes. Later experiments, under conditions more similar to those of the primitive Earth, have confirmed that simple organic compounds can be formed spontaneously. This possibility is now taken for granted, because of a multitude of experiments, but also because simple organic molecules have been found in meteorites falling on our planet, in comets and even in interstellar gas clouds.

The question remains of how these basic building blocks got put together into more complex molecules, such as enzymes and DNA, and into living cells. One favoured scenario is that once the Earth cooled enough to allow oceans to form, something like the processes observed by Miller and others resulted in a broth of organic molecules (a 'primordial soup'), which given enough time (many millions of years were available!) produced by chance combinations of molecules, some of which were more successful than others. At some point a replicating entity would have formed that eventually evolved into life as we know it.

Heredity and mutation

The chicken-and-egg problem still remains. Can DNA or RNA molecules come about, able to direct the synthesis of enzymes that in turn would be able to synthesize DNA and RNA molecules identical, or at least similar, to the pre-existing ones? Could there be continuity of life through the spontaneous synthesis of hereditary molecules that specify the synthesis of the enzymes to carry out the life processes, before there are suitable enzymes? An important advance occurred in the early 1980s when Thomas R. Cech and Sydney Altman independently discovered that some RNA molecules can catalyse chemical reactions, including their own synthesis, a discovery for which they received the Nobel Prize in 1989. This discovery contributed importantly to solving the chicken-and-egg problem, in that these RNA molecules, dubbed ribozymes, can fill the two roles of heredity and metabolism, roles mostly played by separate molecules – DNA and proteins respectively – in current living organisms. Many scientists now believe that life went through an RNA-dominated phase, called the 'RNA world', which preceded the current DNA world, where biological heredity is prevailingly encased in DNA molecules.

Recent efforts have addressed the issue of how ribozyme RNA molecules may have formed spontaneously in the early Earth, eventually leading to the RNA world. Ribozymes, like other RNA molecules, consist of four kinds of nucleotides, represented by the letters A, C, G, and U, as pointed out earlier. They are made up of a limited number of nucleotides, say two dozen or so. But nucleotides are far from simple, containing as they do three molecular components: a ribose sugar, phosphate and a nitrogen base. The nitrogen base is the only constituent that varies from one nucleotide to another. There are four different kinds of nitrogen bases, which correspond to the A, C, G and U nucleotides. Scientists have recently shown how the ribose sugar may become spontaneously linked to the nitrogen bases C and U, which was previously the step most difficult to account for.

The idea that RNA could serve a dual function, as a carrier of information and a catalyst that would carry out the instructions, had already been suggested in the 1960s by Francis Crick, who, with James Watson, had discovered in 1953 the double-helix structure of DNA. The discovery of ribozymes brought attention to Crick's suggestion. The viability of the ribozyme hypothesis is enhanced by consideration of the multiple roles that RNA molecules play in modern organisms. While the role of DNA is largely limited to carrying out the genetic information, RNA is involved in the translation process by which that genetic information is conveyed. As pointed out above, different molecules of messenger RNA, transfer RNA and RNA-protein complexes are involved in the translation process. Small RNA molecules are also involved in DNA splicing and in DNA and RNA processing, and there is a large diversity of RNA molecules (known as RNA interference or RNA*i*) that is extensively involved in all sorts of DNA and RNA signalling.

RNA to life?

Once RNA molecules were formed that could reproduce by copying themselves, albeit subject to some error (mutation) in the synthesis of new RNA molecules, natural selection would occur, gradually leading to greater molecular complexity and eventually to cells: first simpler cells, as in bacteria, and later more advanced ones, as in animals, plants and other eukaryotic organisms. Once there were molecules (RNA) able to reproduce, some were likely to reproduce more effectively than others. And once there were primitive cells able to reproduce, some of these were likely to reproduce more effectively than other cells. The characteristics of the more effectively reproducing cells would increase in frequency at the expense of the less effectively reproducing ones. It stands to reason that the more effectively reproducing cells would often be those that had more precise heredity and more efficient metabolism. The process of evolution, the gigantic project that would give rise to the immense diversity of life forms, had started.

So: do we now know how life began? We don't. What we do know is that in the absence of previous life and under conditions that may plausibly have existed on the early Earth, spontaneous chemical processes can give rise to organic compounds including those that are the fundamental building blocks of life: the nucleic acids that encase heredity and the enzymes that account for metabolism, i.e. the proteins that catalyse the chemical reactions that make up all living processes.

In the absence of previous life and under conditions that may plausibly have existed on the early Earth, spontaneous chemical processes can give rise to organic compounds including those that are the fundamental building blocks of life.

Returning to the transportation network analogy, we now know that early foot trails could be made and how more advanced roads might be developed from them, but we still don't know where those first trails were made or precisely how. What we also know is that life originated on Earth only once; or that if it originated more than once, all forms of life but one became extinct.

Life in the universe

The Earth is probably the only planet in the Solar System that presently has life. There are about 100 billion stars in our galaxy, and many of them have planetary systems. And there are more than 100 billion galaxies in the universe. It may very well be that life exists elsewhere. Some planets, perhaps many of them throughout the universe, may have life, if temperature, chemical composition and other features favourable for life occur, something that seems possible given the immense number of galaxies, stars and planets. Life occurs on Earth because conditions favourable for life exist on our planet. Given similar conditions and aeons of time, the same thing is likely to happen on other planets. Planets with conditions (at least temperature) suitable for life have recently been identified in our galaxy.

Life as it may exist elsewhere would have features different from those listed above, which show the unique origin of life on Earth. The basic chemical elements might be different; for example, silicon rather than carbon might combine with hydrogen, oxygen and other elements to make up the fundamental molecules of life.

WHAT IS THE TREE OF LIFE?
LUCA: Last Universal Common Ancestor

*T*he universal tree of life embraces all living organisms from
their common ancestor to the present. Groups of organisms
are represented by the branches of the tree. There are three major
sets of branches: eukaryotes, bacteria and archaea; the last two
are prokaryotes and are microscopic organisms. Most eukaryotic
organisms are also microscopic single cells, but the familiar animals,
plants and fungi are multicellular organisms. All organisms are
related by common descent from a single form of life: LUCA, the
Last Universal Common Ancestor.

Classification is a necessity. A telephone directory would be
worthless if it listed entries in a random sequence. The benefits of
a public library would be severely curtailed if the books were not
systematically arranged. More than one million species have been
named and described, and many more are known to exist; the
number of extinct species is more than a hundred times the number
of species now in existence. The number of different species that
have lived on Earth from the beginning of life to the present is
likely to be in the order of one billion. The diversity of the living
world is apparent not only in the large number of species, but also
in their heterogeneity. Organisms vary enormously in size, way of
life and habitat, as well as in structure and form.

Despite their prodigious diversity, organisms have much in
common. Certain similarities are shared by some, but not by all.

These similarities can be used to classify, i.e. to characterize, some groups of organisms and distinguish them from others. The basic process responsible for the hierarchy of similarities among living things is evolution – some organisms resemble each other more than others because they are more closely related by lines of descent.

Taxonomy

Some species resemble each other quite closely, and some groups of organisms have features in common so as to give them a sort of family resemblance. Mosquitoes, for example, all look pretty much alike. Although separate groups of insects are rather different from each other when viewed in detail, insects as a whole tend to resemble one another when compared with other distinctive groups of animals, such as worms or clams or fish. There are many more species of insects than of any other comparable group of animals, even though insects are nearly all restricted to the terrestrial habitat; only a relatively few live in truly aquatic environments. In the oceans, the snails

The diversity of the living world is apparent not only in the large number of species, but also in their heterogeneity.

are the group with the most species. Organisms seem to live in every conceivable habitat on the Earth's surface, from the greatest ocean depths to Himalayan mountain peaks, and from sub-zero arctic plains to hot springs. A major reason for the great diversity of living things is the great diversity of environmental conditions found on Earth.

Morphological similarities and differences have been recognized by humans since time immemorial. In classical Greece, Aristotle, and later his followers and those of Plato, notably Porphyry, classified organisms based on their similarities. Taxonomy is the biological discipline that seeks the classification of all organisms. In the eighteenth century, the Swedish naturalist Carolus Linnaeus advanced a system of classification in the tenth edition of his *Systema Naturae* (1758) that was subsequently accepted as the starting point for modern taxonomy.

The rules of classification are slightly different for animals and plants, but the general systems are quite similar. Animal species that resemble each other closely are classed into a genus; genera that resemble each other closely are classed in the same family; families that resemble each other into an order; orders into a class; and classes into a phylum. Phyla are then grouped into a kingdom, such as Metazoa (animals). Plants too are classed in a hierarchical scheme.

Taxonomy is the biological discipline that seeks the classification of all organisms.

Placing species within the hierarchy is only one of the objectives of classification. There must also be some system of identifying species and formal groups of species so that scientific facts about them can be communicated and accumulated in an orderly manner. Popular names of species are not adequate; they vary from language to language, or even from region to region in the same language area. Most species do not have distinctive popular names. Scientists have adopted a special system of nomenclature that identifies each species or group of species unequivocally. The smallest grouping of species is the genus. The complete name of a species consists of the name of the genus followed by the species name itself. For humans, for example, the species name is *Homo sapiens*, and no other animal species may bear this name. The chimpanzee is *Pan troglodytes*, the African leopard is *Felis pardus* and the honeybee is *Apis mellifera*. By convention, the generic name is capitalized and both the generic and the species name are italicized.

The tree of life

The variations on life and the hierarchy of organisms are outcomes of the evolutionary process. Humans are mammals, descended from shrew-like creatures that lived more than 150 million years ago; mammals, birds, reptiles, amphibians and fishes share as ancestors small worm-like creatures that lived in the world's oceans 600 million years ago; plants and animals are derived from bacteria-like microorganisms that originated more than three billion years ago. Because of biological evolution, lineages of organisms change through time; diversity arises because lineages that descend from common ancestors diverge through the generations as they become adapted to different ways of life. The more recent the last common ancestor of two species is, the more similar the two species are likely to be, since evolution is a continuing process of biological change. Hence, biologists have found it convenient to represent the gradual succession of species through time in a tree-like configuration. Darwin's only figure in *On the Origin of Species* is a branching diagram in Chapter 4 to illustrate how the descendants of one species would gradually diverge over time. An explicit tree-like figure encompassing the whole of life was first published in 1866 by the German naturalist Ernst Haeckel.

There are three large groups of organisms on Earth: eukaryotes, bacteria and archaea. Eukaryotes are organisms that have their genetic material enclosed in a special capsule, or organelle, called the nucleus. We are eukaryotes. Animals, plants and fungi are the only organisms large enough to be experienced directly with our senses, and thus up to three centuries ago they were the only organisms whose existence was known to humans. Yet they account for only a fraction of the total diversity of the eukaryotes. The other eukaryotes are all microscopic. Some cause well-known diseases; *Plasmodium*, for instance, causes malaria, while *Entamoeba* is responsible for severe intestinal maladies.

The more recent the last common ancestor of two species is, the more similar the two species are likely to be, since evolution is a continuing process of biological change.

Humans have known of the existence of bacteria, the second group of organisms, for more than a century. We associate them with diseases, but bacteria perform many useful functions, including the incorporation from the atmosphere of nitrogen, which animals and plants need but are not able to get directly (it is very abundant in the atmosphere, about 75 per cent of the total; the rest is mostly oxygen). Bacteria are also responsible for the decomposition of dead matter, a process that is essential in the maintenance of the cycle of life and death, because it makes available for new organisms valuable components from dead ones.

The genetic diversity and number of species of bacteria is huge. There are many more kinds of bacteria than there are kinds of animals, plants and fungi combined. And they are so abundant that their total weight (biomass) is at least as great as (and probably much greater than) that of all plants, fungi and animals put together, even though individually they are so much smaller. This is a humbling thought. We see ourselves, the human species, as the summit of life, as well as the most numerous of all large animals; and we see animals and plants as the dominant forms of life on Earth. However, modern biology teaches us that, numerically as well as in biomass, the one million known species of animals (including humans) amount to only a very small fraction of life on Earth.

The third group, the archaea, is likely to be about as large as the eukaryotes or the bacteria. The existence of the archaea is a very recent discovery of molecular biology. Because these organisms do not interact much with us directly, we were not aware of their existence. Biologists knew a few species, such as those that exist in the hot springs of Yellowstone National Park in the United States, and in other volcanic hot springs, where they thrive at temperatures approaching the boiling point of water. It was thought that these were some unusual forms of bacteria. Now we know that they belong to a very diverse and numerous group of organisms, abundant in the top water layers of the seas and oceans. A bucket of seawater studied with the modern techniques of molecular biology may yield tens or hundreds of new archaea species.

Fossils and forms

Evolutionists seek to discover the lines of descent of organisms in order to reconstruct the ancestral-descendant pathways among living and extinct organisms. The objective is to determine the genealogy of life, which is expressed in family trees that depict the pattern of relatedness among different groups of organisms, at different levels of resolution and scope. Trees may seek to resolve the phylogenetic relationships among closely related species, say, the Galápagos finches observed by Darwin; or among much more distantly related organisms, such as the set of amphibians, reptiles, birds and mammals, or even among the whole of life – i.e. the universal tree of life – something that became possible only a few decades ago, thanks to molecular biology.

One way to reconstruct phylogenies consists of tracing a series of fossils that resemble each other but show a sequence of changes leading through time from an ancestral to a descendant form. Relationships among the fossils are thus judged by their relative ages and their morphological resemblances and differences. This works well when abundant fossils are available in a continuous record, but unfortunately the fossil record is quite incomplete. Most animals do not have easily fossilizable hard parts, and only a small fraction of animals with shells or bones are actually preserved as fossils. Other organisms, from plants, fungi and protozoa to archaea and bacteria, have little or no fossil record at all. For most lineages we have to employ other methods of phylogenetic reconstruction.

Genes and proteins

The discipline of molecular biology came into existence in the second half of the twentieth century. Molecular biology has become the most effective way of reconstructing evolutionary phylogeny because, as pointed out in *What is Molecular Evolution?*, it has three advantages over the traditional methods of classification based on fossils or comparative morphology: precision, because the information is readily quantifiable; universality, because all sorts of organisms, no matter how evolutionarily distant, can be compared; and multiplicity, because organisms have thousands of genes and proteins, every one of which reflects the same evolutionary history; more and more genes and proteins can be investigated in order to achieve the desired degree of resolution in the tree.

The Universal Tree of Life represented in the figure on page 105 can only be reconstructed with the methods of molecular biology. The tree in the figure was obtained with the genes that encode ribosomal RNA molecules, because they are among the slowest evolving genes, meaning that different degrees of similarity can be detected even among the most diverse organisms, such as bacteria compared to archaea and to eukaryotes.

Distance methods

If we need to resolve the evolutionary history of particular organisms using molecular methods, the first step is to select a gene or protein and to align the nucleotides in DNA, or the amino acids in proteins, among the species to be assessed (see page 85). There are several techniques for constructing molecular evolutionary trees based on the differences observed. Some approaches were first developed for interpreting palaeontological (fossils) or morphological data; others for interpreting molecular data. The main methods currently in use are called distance, parsimony and maximum likelihood.

A distance is the number of differences between two taxa. (A taxon – plural taxa – is a group of related organisms.) The rate of evolution is often assumed to be constant through time for all organisms, so that the number of differences are proportional to the time elapsed. The differences are measured with respect to the sequence of amino acids in proteins or nucleotides in DNA (or RNA). The first step is to obtain a table that includes the number of differences between all pairs of species. The method known as 'cluster analysis' will first identify the two taxa with the smallest number of differences and join them as two branches. The next smallest distance between two species is then identified and these two species are similarly joined to one another. When a new distance involves a taxon that is already included in a cluster, the average distance is obtained between the new taxon and the pre-existing cluster, and so on until all taxa have been included in the branching tree.

The relationships between species based on cytochrome c shown in the figure on page 86 correspond fairly well to those determined from other sources, such as the fossil record. According to the figures, chickens are less closely related to ducks and pigeons than to penguins, and humans and monkeys

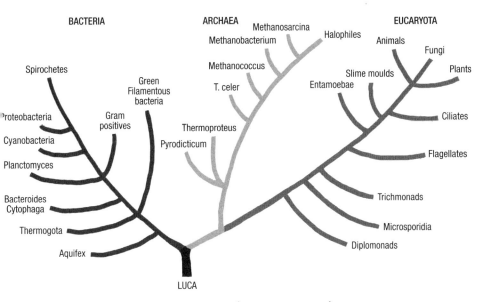

THE UNIVERSAL TREE OF LIFE, RECONSTRUCTED WITH rRNA (RIBOSOMAL NUCLEIC ACID) GENES. THE THREE MAJOR GROUPS OF ORGANISMS – BACTERIA, ARCHAEA, AND EUCARYOTA – THAT HAVE ALL EVOLVED FROM A COMMON ANCESTOR, LUCA.

diverged from the other mammals before the marsupial kangaroo separated from the non-primate placentals. These are known to be erroneous relationships, but the power of the method is apparent in that a single protein yields a fairly accurate reconstruction of the evolutionary history of 20 organisms that started to diverge as early as one billion years ago.

Some distance methods (including the one used to construct the tree on page 86) relax the condition of uniform rate of evolution and allow for unequal rates of evolution along the branches. One of the most extensively used methods of this kind is called neighbour-joining. The method starts, as before, by identifying the smallest distance in the table and linking the two taxa involved. The next step is to remove these two taxa and calculate a new table in which their distances to other taxa are replaced by the distance between the node linking the two taxa and all other taxa. The smallest distance in this new table is used to make the next connection, which will be between two other taxa or between the previous node and a new taxon. The procedure is repeated until all taxa have been connected with one another by intervening nodes.

Parsimony methods

Maximum parsimony methods seek to reconstruct the tree that requires the fewest (i.e. the most parsimonious) number of changes summed along all branches. This is reasonable, because it will be the most likely if no other information is available. But evolution may not necessarily have occurred following a minimum path. The same change may have occurred independently along different branches, and some changes may have involved intermediate steps or back mutations.

Maximum parsimony methods are related to cladistics, a very formalistic theory of taxonomic classification, extensively used with morphological and palaeontological data. The critical feature in cladistics is the identification of derived shared traits, called synapomorphic traits. A synapomorphic trait is shared by some taxa but not others because the former inherited it from a common ancestor that acquired the trait after its lineage separated from the lineages going to the other taxa. In the evolution of carnivores, for example, domestic cats, tigers and leopards are clustered together because they possess retractable claws, a trait acquired after their common ancestor branched off from the lineage leading to the dogs, wolves and coyotes. It is important to ascertain that the shared traits are homologous rather than analogous. For example, mammals and birds, but not lizards, have a four-chambered heart. Yet birds are more closely related to lizards than to mammals; the four-chambered heart evolved independently in the bird and mammal lineages, by parallel evolution.

Maximum likelihood

Maximum likelihood methods seek to identify the most likely tree, given the available data. They require that an evolutionary model be identified that would make it possible to estimate the probability of each potential individual change. All possible trees are considered. The probabilities for each individual change are multiplied for each tree. The best tree is the one with the highest probability (or maximum likelihood) among all the trees.

Maximum likelihood methods are computationally expensive when the number of taxa is large, because the number of possible trees (for each of which the probability must be calculated) grows factorially with the number of taxa. With 10 taxa, there are about 3.6 million possible trees; with 20 taxa, the number of possible trees is about 2 followed by 18 zeros (2×10^{18}). Even with powerful computers, maximum likelihood methods can be prohibitive if the number of taxa is large. Heuristic methods exist in which only a subsample of all possible trees is examined and thus an exhaustive search is avoided.

The statistical degree of confidence of a tree can be estimated for distance and maximum likelihood trees. The most common method is called bootstrapping. It consists of taking samples of the data by removing at least one data point at random and then constructing a tree for the new data set. This random sampling process is repeated hundreds or thousands of times. The bootstrap value for each node is defined by the percentage of cases in which all species derived from that node appear together in the trees. Bootstrap values above 90 per cent are regarded as statistically strongly reliable; those below 70 per cent are considered unreliable.

AM I REALLY A MONKEY?
Biology to culture

*H*umans and chimpanzees shared common ancestors as recently *as six or seven million years ago. Humans, apes and monkeys are primates, but humans are a very special kind of primate. We have bipedal gait, large brains and opposable thumbs that make possible precise manipulation of tools and other objects. Moreover, and most important, humans have superior intelligence and culture, which includes language, advanced technology and complex social and political institutions, as well as ethics and religion.*

It does not take a great deal of biological expertise to realize that humans have organs and limbs similar to those of other animals; that we bear our young like other mammals; that bone by bone there is a precise correspondence between the skeletons of a chimpanzee and a human. But it also does not take much reflection to notice the distinct uniqueness of our species. Conspicuous anatomical differences include bipedal gait and enlarged brain. Much more noticeable than the anatomical differences are the distinct behaviours. Humans have elaborate social and political institutions, codes of law, literature and art, ethics and religion; we build roads and cities, travel by motorcar, ship and aeroplane and communicate by means of telephones, computers and televisions.

Primates and hominids
Monkeys, apes and humans all belong to the category of mammals known as primates. That is, the apes are our first cousins, so to

speak, while the monkeys are our second or third cousins. Our last common ancestors with monkeys lived about 25–30 million years ago, with chimpanzees about six or seven million years ago, with gorillas about eight to ten million years ago, and with orang-utans more than ten million years ago. We know about these matters in three ways: by comparing living primates, including humans, with each other; by the discovery and investigation of fossil remains of primates that lived in the past; and by comparing their DNA, proteins and other molecules. DNA and proteins give us the best information about how closely related we are to each of the primates, and those to each other. But in order to know how the human lineage changed in anatomy and behaviour over time as our ancestors became more and more human-like, we have to study fossils and the tools and other remnants of their activities. This information is reviewed in *What is the Missing Link?*.

Our closest biological relatives are the great apes, and among them the chimpanzees, who are more closely related to us than they are to the gorillas, and much more than to the orang-utans. The hominid lineage diverged from the chimpanzee lineage 6–7 million years ago (Ma), evolving exclusively in the African continent until the appearance of *Homo erectus*, somewhat before 1.8 Ma. Shortly after its emergence in tropical or subtropical eastern Africa, *H. erectus* spread to other continents. Fossil remains of *H. erectus* (*sensu lato*) are known from Africa, Indonesia (Java), China, the Middle East and Europe. *H. erectus* fossils from Java have been dated 1.81±0.04 and 1.66±0.04 Ma, and from Georgia 1.6–1.8 Ma. Anatomically distinctive *H. erectus* fossils have been found in Spain, deposited before 780,000 years ago, the oldest in southern Europe.

The transition from *H. erectus* to *H. sapiens* occurred around 400,000 years ago, although this date is not well determined owing to uncertainty as to whether some fossils are *erectus* or 'archaic' forms of *sapiens*. *H. erectus* persisted for some time in Asia, until 250,000 years ago in China and perhaps until 100,000 years ago in Java, and thus was coetaneous with early members of its descendant species, *H. sapiens*. Fossil remains of Neanderthal hominids (*Homo neanderthalensis*), with brains as large as those of *H. sapiens*, appeared

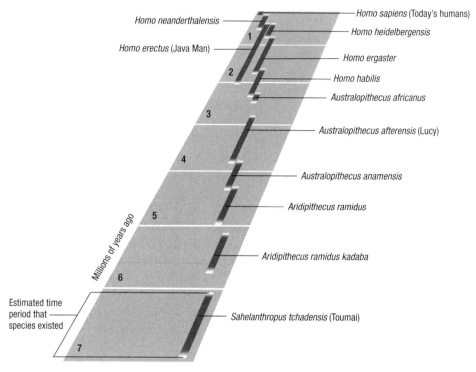

Homo sapiens (Today's humans)

Homo neanderthalensis

Homo heidelbergensis

Homo erectus (Java Man)

Homo ergaster

Homo habilis

Australopithecus africanus

Australopithecus afterensis (Lucy)

Australopithecus anamensis

Aridipithecus ramidus

Millions of years ago

Aridipithecus ramidus kadaba

Estimated time period that species existed

Sahelanthropus tchadensis (Toumai)

HOMINID SPECIES, SHOWING SEVERAL INTERMEDIATE SPECIES BETWEEN *SAHELANTHROPUS TCHADENSIS* (BOTTOM), WHICH LIVED 6–7 MILLION YEARS AGO AND MODERN HUMANS AT THE TOP.

in Europe around 200,000 years ago and persisted until 30,000 or 40,000 years ago.

There is controversy about the origin of modern humans. Some anthropologists argue that the transition from *H. erectus* to archaic *H. sapiens* and later to anatomically modern humans occurred consonantly in various parts of the Old World. Proponents of this 'multiregional model' emphasize fossil evidence showing regional continuity in the transition. Others argue instead that modern humans first arose in Africa or in the Middle East somewhat prior to 100,000 years ago and from there spread throughout the world, replacing elsewhere the pre-existing populations of *H. erectus* or archaic *H. sapiens*. The African (or Middle East) origin of modern humans is supported by a wealth of recent genetic evidence and is therefore favoured by many evolutionists.

Some proponents of this 'African replacement' model claim further that the transition from archaic to modern *H. sapiens* was associated with a very narrow bottleneck consisting of only two or very few individuals who are the ancestors of all modern mankind. This particular claim of a narrow bottleneck is supported, erroneously, by the investigation of a peculiar small fraction of our genetic inheritance, the mitochondrial DNA (mtDNA).

The myth of the mitochondrial Eve

The argument that modern humans descend from only one or very few ancestral women is based on the analysis of the mitochondrial DNA from ethnically diverse individuals. Phylogenies of the mtDNA sequences of modern humans coalesce to one ancestral sequence, the 'mitochondrial Eve' that existed in Africa about 200,000 years ago. This Eve, however, is not the one mother from whom all humans descend, but an mtDNA molecule (or the female carrier of that molecule) from whom all modern mtDNA molecules descend. The mtDNA is inherited only from the mother, who transmits it to sons as well as daughters, but only the daughters transmit it to their own children.

This Eve is not the one mother from whom all humans descend, but an mtDNA molecule (or the female carrier of that molecule) from whom all modern mtDNA molecules descend.

The myth of the mitochondrial Eve is based on a confusion between gene genealogies (phylogenies) and individual genealogies. Gene genealogies gradually coalesce towards a unique DNA ancestral sequence (in a similar fashion as living species, such as humans, chimpanzees and gorillas, coalesce into one ancestral species). Individual genealogies, on the other hand, increase by a factor of two in each ancestral generation: an individual has two parents, four grandparents and so on. Coalescence of a gene genealogy into one ancestral gene, originally present in one individual, does not disallow the contemporary existence of many other individuals, who are also our ancestors, and from whom we have inherited the other genes.

This conclusion can be illustrated with an analogy. My family name, Ayala, is shared by many people in Spain, Mexico, the Philippines and other countries. A historian of our family name has concluded that all Ayalas descend from Don Lope Sánchez de Ayala, grandson of Don Vela, vassal of King Alfonso VI, who established the domain (*señorío*) de Ayala in the year 1085, in the now Spanish Basque province of Alava. Don Lope is the Adam from whom we all descend on the paternal line, but we also descend from many other men and women who lived in the eleventh century, as well as earlier and later.

The inference warranted by the mtDNA analysis is that the so-called mitochondrial Eve is the ancestor of modern humans in the maternal line. Any person has a single ancestor in the maternal line in any given generation. Thus a person inherits the mtDNA from the mother, from the maternal grandmother, from the great-grandmother on the maternal line and so on. But they also inherit other genes from other ancestors. The mtDNA that we have inherited from the mitochondrial Eve represents one four-hundred-thousandth of the DNA present in any modern human (16,000 out of 6 billion nucleotides). The rest of the DNA, 400,000 times more than the mtDNA, we have inherited from other contemporaries of the mitochondrial Eve.

The ape-to-human transformation

By the 'ape-to-human transformation' I mean the mystery of how a particular ape lineage became a hominid lineage, from which emerged, after only a few million years, humans able to think and love, who have developed complex societies and who uphold ethical, aesthetic and religious values.

The ape-to-human transformation is a major concern for many people. Are scientists claiming that humans are just another kind of ape, no more different from chimpanzees than gorillas and other apes are? Does this imply that the view of humans as very special creatures is without foundation? The answer to these questions is that in some biological respects we are very similar to apes, but in others we are very different, and these differences provide a valid foundation for a view of humans as very special creatures.

The Human Genome Project

How are humans and apes similar and how do they differ? Biological heredity in both humans and other animals is based on the transmission of genetic information from parents to offspring. The DNA of humans is packaged in two sets of 23 chromosomes, one set inherited from each parent. The total number of DNA letters (the four nucleotides represented by A, C, G and T) in each set of chromosomes is about three billion. The Human Genome Project has deciphered the sequence of the three billion letters in a human genome (i.e. in one particular set of chromosomes; the human genome sequence varies between genomes by about one letter in a thousand).

The two genomes (chromosome sets) of each individual are different from each other, and from the genomes of any other human being (with the trivial exception of identical twins, who share the same two sets of genes because identical twins develop from one single fertilized human egg). The King James Bible contains an estimated three million letters, punctuation marks and spaces. Writing down the DNA sequence of one human genome would require 1,000 volumes the size of the Bible. The human genome sequence is, of course, not printed in books, but stored in electronic form on computers from which fragments of information can be retrieved by investigators. But if a printout is wanted, 2,000 volumes will be needed for just one individual, 1,000 for each of the two chromosome sets. Surely, though, there are more economic ways of presenting the information in the second set than listing the complete letter sequence; for example, by indicating the position of each variant letter in the second set relative to the first. The number of variant letters between one individual's two sets is about three million, or 1 in 1,000.

The Human Genome Project of the United States was initiated in 1989, funded through two agencies, the National Institutes of Health and the Department of Energy. (A private enterprise, Celera Genomics, started in the United States somewhat later, but achieved, largely independently, similar results.) The goal was the complete sequence of one human genome in 15 years at an approximate cost of $3 billion, coincidentally about one dollar per DNA letter. A draft of the genome sequence was completed ahead of schedule in 2001. In 2003, the Human Genome Project was finished.

Knowing the human DNA sequence is a first step, but no more than that, towards understanding the genetic make-up of a human being. Think of the 1,000 Bible-sized volumes. We now know the orderly sequence of the three billion letters, but this sequence does not provide an understanding of human beings any more than we would understand the contents of 1,000 Bible-sized volumes written in an extraterrestrial language, of which we only know the alphabet, just because we have come to decipher their letter sequence.

The chimpanzee genome

A draft of the DNA sequence of the chimpanzee genome was published on 1 September 2005. In the genome regions shared by humans and chimpanzees, the two species are 98–99 per cent identical. The differences appear to be either very small or quite large, depending on how one chooses to look at them: 1 per cent of the total sounds very little, but it amounts to a difference of 30 million DNA letters out of the 3 billion in each genome. Twenty-nine per cent of the enzymes and other proteins encoded by the genes are identical in both species. Out of the one hundred to several hundred amino acids that make up each protein, the 71 per cent of non-identical proteins differ between humans and chimps by only two amino acids on average. If one takes into account DNA segments found in one species but not the other, the two genomes are about 96 per cent identical, rather than nearly 98–99 per cent identical as in the case of DNA sequences shared by both species. That is, a large amount of genetic material – about 3 per cent, or some 90 million DNA letters – has been inserted or deleted since humans and chimps initiated their separate evolutionary ways, six to seven million years ago. Most of this DNA does not contain genes coding for proteins.

Comparison of the two genomes provides insights into the rate of evolution of particular genes in the two species. One significant finding is that genes active in the brain have changed more in the human lineage than in the chimp lineage. Also significant is that the fastest evolving human genes are those coding for 'transcription factors', that is, 'switch' proteins, which control the expression of other genes, determining when they are turned on and off. On the whole, 585 genes, including genes involved in resistance to malaria and tuberculosis, have been identified as evolving faster in humans than in chimps. (Note that malaria is a severe disease for humans but not for chimps.)

Genes located on the Y chromosome, found only in the male, have been much better protected by natural selection in the human than in the chimpanzee lineage, in which several genes have incorporated disabling mutations that make the genes non-functional. Also, several regions of the human genome seem to contain beneficial genes that have rapidly evolved within the past 250,000 years. One of these is the *FOXP2* gene, involved in the evolution of speech.

> *In an important sense, the most distinctive human features are those expressed in the brain, those that account for the human mind and for human identity.*

Extended comparisons of the human and chimp genomes and experimental exploration of the functions associated with significant genes will surely advance considerably our understanding, over the next decade or two, of what it is that makes us distinctively human. The features that distinguish us as human begin early in development, well before birth, as the linear information encoded in the genome gradually becomes expressed into a four-dimensional individual, an individual who changes in configuration as time goes by. In an important sense, the most distinctive human features are those expressed in the brain, those that account for the human mind and for human identity.

Human uniqueness

The most distinctive human anatomical traits are erect posture and large brain. We are the only vertebrate species with a bipedal gait and erect posture; birds are bipedal, but their backbone stands more nearly horizontal than vertical. Brain size is generally proportional to body size; relative to body mass, humans have the largest (and most complex) brain. The human male adult brain is 1,400 cubic centimetres (cc) and about three pounds in weight, three times as large as the chimpanzee brain.

Evolutionists used to raise the question whether bipedal gait or large brain came first in the evolution of the human lineage, or whether they evolved consonantly. The issue is now resolved. Our ancestors had, more than four million years ago, a bipedal gait, but those ancestors possessed a small brain, about 450 cc and a pound in weight. Brain size started to increase notably with our *Homo habilis* ancestors, 2–2.5 Ma, who had brains of 650–800 cc. *Homo erectus* had adult brains of about 1,200 cc. As mentioned above, our species, *Homo sapiens*, has a brain of about 1,300–1,400 cc, or some three pounds of grey matter.

Erect posture and large brain are not the only anatomical traits that distinguish us from non-human primates. A list of our most distinctive anatomical features includes:

- erect posture and bipedal gait (entail changes of the backbone, hipbone and feet)
- opposable thumbs and arm and hand changes (make possible precise manipulation)
- large brain
- reduction of jaws and remodelling of face
- changes in skin and skin glands
- reduction in body hair
- cryptic ovulation (and continuous female sexual receptivity)
- slow development
- modification of vocal tract and larynx
- reorganization of the brain

Humans are notably different from other animals not only in anatomy, but also and no less importantly in their behaviour, both as individuals and socially. Distinctive human behavioural traits include the following:

- subtle expression of emotions
- intelligence: abstract thinking, categorizing and reasoning
- symbolic (creative) language
- self-awareness and death awareness
- tool-making and technology
- science, literature and art
- ethics and religion
- social organization and cooperation
- legal codes and political institutions

Biological evolution and cultural evolution

Both humans and other primates live in groups that are socially organized. But primate societies do not approach the complexity of human social organization. A distinctive human social trait is culture, which may be understood as the set of non-strictly biological human activities and creations. Culture includes social and political institutions, religious and ethical traditions, language, common sense and scientific knowledge, art and literature, technology and in general all the creations of the human mind. The advent of culture has brought with it cultural evolution, a superorganic mode of evolution superimposed on the organic mode that has become dominant. Cultural evolution has come about because of cultural change and inheritance, a distinctively human means of achieving adaptation to the environment.

There are in mankind two kinds of heredity – the biological and the cultural, which may also be called organic and superorganic systems of heredity. Biological inheritance in humans is very much like that in any other sexually reproducing organism; it is based on the transmission of genetic information encoded in DNA from one generation to the next by means of the sex cells. Cultural inheritance, on the other hand, is based on transmission of information by a teaching-learning process, which is in principle independent of biological parentage. Culture is transmitted by instruction and learning, by example and imitation, through books, newspapers and radio, television and motion pictures, through works of art and by any other means of communication. Culture is acquired by every person from parents, relatives and neighbours, and from the whole human environment.

Cultural inheritance makes possible for people what no other organism can accomplish – the cumulative transmission of experience from generation to generation. Animals can learn from experience, but they do not transmit their experiences, their 'discoveries' (at least not to any large extent) to the following generations. Animals have individual memory, but they do not have a 'social memory'. Humans, on the other hand, have developed a culture because they can transmit cumulatively their experiences from generation to generation.

Cultural inheritance makes possible for people what no other organism can accomplish – the cumulative transmission of experience from generation to generation.

Cultural inheritance enables cultural evolution, that is, the evolution of knowledge, social structures, ethics and all other components that make up human culture. It also makes possible a new mode of adaptation to the environment that is not available to non-human organisms – adaptation by means of culture. Organisms in general adapt to the environment through natural selection, by changing over generations their genetic constitution to suit the demands of the environment. But humans, and humans alone, can also adapt by changing the environment to suit the needs of their genes. (Animals build nests and modify their environment in other ways, but the manipulation of the environment by any non-human species is trivial compared to mankind's.) For the last few millennia humans have been adapting the environment to their genes more often than their genes to the environment.

In order to extend its geographical habitat, or to survive in a changing environment, a population of organisms must become adapted, through slow accumulation of genetic variants sorted out by natural selection, to the new climatic conditions, sources of food, competitors and so on. The discovery of fire and the use of shelter and clothing allowed humans to spread from the warm tropical and subtropical regions of the Old World to the whole of the Earth, except for the frozen wastes of Antarctica, without the anatomical development of fur or hair. Humans did not wait for genetic mutants promoting wing development; they have conquered the air in a somewhat more efficient and versatile way by building flying machines. People travel the rivers and the seas without gills or fins. The exploration of outer space has started without waiting for mutations providing us with the ability to breathe with low oxygen pressures or to function in the absence of gravity; astronauts carry their own oxygen and specially equipped pressure suits. From their obscure beginnings in Africa, humans have become the most widespread and abundant species of mammal on earth. It was the appearance of culture as a superorganic form of adaptation that allowed this to happen.

Cultural adaptation has prevailed in mankind over biological adaptation because it is a more effective mode; it is more rapid and it can be directed. A favourable genetic mutation newly arisen in an individual can be transmitted to a sizeable part of the human species only through innumerable generations. However, a new scientific discovery or technical achievement can be transmitted to the whole of mankind, potentially at least, in less than one generation. Witness the rapid spread of personal computers, iPhones and the internet. Moreover, whenever a need arises, culture can directly pursue the appropriate changes to meet the challenge. Biological adaptation, on the other hand, depends on the accidental availability of a favourable mutation, or of a combination of several mutations, at the time and place where the need arises.

WHAT DOES THE FOSSIL RECORD TELL US?

Life has existed on Earth for most of our planet's history

'*Palaeontologists have recovered and studied the fossil remains of many thousands of organisms that lived in the past. These fossils show that many kinds of extinct organisms were very different in form from any now living. The fossil record also shows successions of organisms through time, as well as their transition from one form to another.*'[4]

Palaeontology was a rudimentary science in Darwin's time, and large parts of the geological succession of stratified rocks were unknown or inadequately studied. Darwin, therefore, worried about the rarity of intermediate forms between major groups of organisms. Anti-evolutionists then and now have seized on this as a weakness in evolutionary theory. Although gaps in the palaeontological record remain even today, many have been filled by the researches of palaeontologists since Darwin's time. Hundreds of thousands of fossil organisms found in well-dated rock sequences represent a succession of forms through time and manifest many evolutionary transitions.

Age of the Earth and life

When an organism dies, it is usually destroyed by bacteria and other organisms and by weathering processes. On rare occasions some body parts – particularly hard ones such as shells, teeth and bones – are buried in mud or protected in some other way from predators, decomposition and weather, and these can be preserved indefinitely within the rocks in which they are embedded. (Mud and other sediments may over time become limestone and other sorts of rocks.) Methods such as radiometric dating – measuring the amounts of radioactive atoms that remain in certain minerals – make it possible to estimate the time period when the rocks, and the fossils associated with them, were formed.

Radiometric dating indicates that Earth was formed about 4.5 billion years ago. The earliest known fossils were recently discovered, and date to around 3.4 billion years ago. The oldest known animal fossils, nearly 700 million years old, come from the Ediacara fauna, small wormlike creatures with soft bodies. Numerous fossils belonging to many animal phyla and exhibiting mineralized skeletons appear in rocks about 540 million years old, during the geological period known as the Cambrian. (A phylum, plural phyla, is a major group of organisms, such as the molluscs or the chordates.) These organisms are different from those living now and from those living at intervening times. Some are so radically different that palaeontologists have had to create new phyla in order to classify them. The first vertebrates (phylum Chordata, or chordates) appeared more than 400 million years ago; the first mammals less than 200 million years ago. The history of life recorded by fossils presents compelling evidence of evolution.

Microbial life of the simplest type (i.e. prokaryotes, cells whose nuclear matter is not bounded by a nuclear membrane) was already in existence more than 3 billion years ago. The oldest evidence suggesting the existence of more complex organisms (i.e. eukaryotic cells with a true nucleus) has been discovered in fossils that had been sealed in flinty rocks approximately 1.4 billion years old, but such complex cells are thought to have originated as early as 2,000 million years ago. More advanced forms like true algae, fungi, higher plants and animals have been found only in younger geological strata. The following list presents the order in which progressively complex forms of life appeared:

LIFE FORM	MILLIONS OF YEARS SINCE FIRST KNOWN APPEARANCE
Microbial (prokaryotic cells)	3,500
Complex (eukaryotic cells)	1,400–2,000
First multicellular animals	670
Shell-bearing animals	540
Vertebrates (simple fishes)	490
Amphibians	350
Reptiles	310
Mammals	200
Non-human primates	60
Earliest apes	25
Australopithecine ancestors	5
Homo sapiens (modern humans)	0.15 (150,000 years)

The sequence of observed forms and the fact that all except the first on the list (prokaryotes) are constructed from the same basic cellular type strongly implies that all these major categories of life (including animals, plants and fungi) have a common ancestry in the first eukaryotic cell. Moreover, there have been so many discoveries of intermediate forms between fish and amphibians, between amphibians and reptiles, between reptiles and mammals, and even along the primate line of descent from apes to humans that it is often difficult to identify categorically where the transition occurs from one to another particular genus or from one to another particular species. Nearly all fossils can be regarded as intermediates in some sense; they are life forms that come between ancestral forms that preceded them and those that followed.

The fossil record thus provides compelling evidence of systematic change through time – descent with modification.

From this consistent body of evidence it can be predicted that no reversals will be found in future palaeontological studies. That is, amphibians will not appear before fishes nor mammals before reptiles, and no complex life will occur in the geological record before the oldest eukaryotic cells.

Although some creationists have claimed that the geological record, with its orderly succession of fossils, is the product of a single universal flood a few thousand years ago that lasted a little longer than a year and covered the highest mountains to a depth of some seven metres, there is clear evidence in the form of inter-tidal and terrestrial deposits that at no recorded time in the past has the entire planet been under water. Moreover, a universal flood of sufficient magnitude to deposit the existing strata, which together are many scores of kilometres thick, would require a volume of water far greater than has ever existed on and in the Earth, at least since the formation of the first known solid crust about four billion years ago. The belief that all this sediment with its fossils was deposited in an orderly sequence in only a year defies all geological observations and physical principles. There were periods of unusually high rainfall, and extensive flooding of inhabited areas has occurred, but there is no scientific support for the hypothesis of a universal mountain-topping flood.

The fossil record thus provides compelling evidence of systematic change through time – descent with modification.

Fossils

Although it was Darwin, above all others, who first marshalled convincing evidence for biological evolution, earlier scholars had recognized that organisms on Earth had changed systematically over long periods of time. For example, in 1799, an engineer named William Smith reported that, in undisrupted layers of rock, fossils occurred in a definite sequential order, with more modern-appearing ones closer to the top. Because bottom layers of rock logically were laid down earlier and thus are older than top layers, the sequence of fossils also could be given a chronology from oldest to youngest. Smith's findings were confirmed and extended in the 1830s by the palaeontologist William Lonsdale, who recognized that fossil remains of organisms from lower strata were more primitive than the ones above. Georges Cuvier is often considered the founding father of palaeontology. As a member of the faculty at the National Museum of Natural Sciences in Paris in the early nineteenth century, he had access to the most extensive collection of fossils available at the time. He was adamant that fossils of animals different from those now living represented extinct species rather than ancestors of current species. The older contemporary of Darwin, Sir Charles Lyell, in his *Principles of Geology* (1830–3), proposed that Earth's physical features were the outcome of major geological processes acting over immense periods of time, incomparably greater than the few thousand years since Creation that was commonly believed at the time. Today, many thousands of ancient rock deposits have been identified that show corresponding successions of fossil organisms.

> *Of the small proportion of organisms preserved as fossils, only a tiny fraction have been recovered and studied by palaeontologists*

Thus, the general sequence of fossils had already been recognized before Darwin conceived of his theory of descent with modification, or evolution by natural selection as we would more usually say. The palaeontologists and geologists before Darwin (including Charles Lyell) used the sequence of fossils in rocks not as proof of biological evolution, but mostly for working out the original sequence of rock strata that had been structurally disturbed by earthquakes and other forces. Nevertheless, in Darwin's time, palaeontology was still a rudimentary science. Large parts of the geological succession of stratified rocks were unknown or inadequately studied.

The fossil record is still incomplete. Of the small proportion of organisms preserved as fossils, only a tiny fraction have been recovered and studied by palaeontologists, although in numerous cases the succession of forms over time has been reconstructed in considerable detail. One example is the evolution of the horse, which can be traced to an animal the size of a dog with several toes on each foot and teeth appropriate for browsing (eating tender shoots, twigs and leaves of trees and shrubs); this animal, called the dawn horse (scientific name *Hyracotherium*), lived more than 50 million years ago. The most recent form, the modern horse (*Equus*), is much larger, is one-toed and has teeth appropriate for grazing (eating growing herbage). Transitional forms are well preserved as fossils, as are other kinds of extinct horses that evolved in different directions and left no living descendants (see *Is Evolution a Random Process?*).

Using fossils, palaeontologists have reconstructed examples of radical evolutionary transitions in form and function. For example, the lower jaw of reptiles consists of several bones, but that of mammals has only one. The other bones in the reptilian jaw evolved into bones now found in the mammalian ear. At first, such a transition would seem unlikely – it is hard to imagine what function such bones could have had during their intermediate stages. Yet palaeontologists have discovered two transitional forms of mammal-like reptiles, called therapsids, that had a double jaw joint (i.e. two hinge points side by side) – one joint consisting of the bones that persist in the mammalian jaw and the other composed of the quadrate and articular bones, which eventually became the hammer and anvil of the mammalian ear.

Archaeopteryx and *Tiktaalik*

Many fossils intermediate between diverse organisms have been discovered over the years. Two examples that have received recent attention in the media are *Archaeopteryx*, an animal intermediate between reptiles and birds, and *Tiktaalik*, intermediate between fishes and tetrapods (animals with four limbs).

The first *Archaeopteryx* was discovered in Bavaria in 1861, two years after the publication of Darwin's *Origin,* and received much attention because it shed light on the origin of birds and bolstered Darwin's postulate of the existence of missing links. Other *Archaeopteryx* specimens have been discovered in the past hundred years. The most recent, the tenth specimen so far recovered, was described in December 2005. The best preserved *Archaeopteryx* yet, it is now housed in a small, privately owned museum in Thermopolis, Wyoming. The tetrapod-like fish *Tiktaalik* is also a very recent discovery, published only on 6 April 2006.

Archaeopteryx lived during the Late Jurassic period, about 150 million years ago, and exhibited a mixture of both avian and reptilian traits. All known specimens are small, about the size of a crow, and share many anatomical characteristics with some of the smaller bipedal dinosaurs. The skeleton is reptile-like, but *Archaeopteryx* had feathers, clearly shown in the fossils, with a skull and a beak like those of a bird.

Palaeontologists have known for more than a century that tetrapods (amphibians, reptiles, birds and mammals) evolved from a particular group of fishes called lobe-finned. Until recently, *Panderichthys* was the closest known fossil fish to the tetrapods. It was somewhat crocodile-shaped and had a pectoral fin skeleton and shoulder girdle intermediate in shape between those of typical lobe-finned fishes and those of tetrapods, which allowed it to 'walk' in shallow waters, but probably not on land. In most features, however, it was more like a fish than a tetrapod. *Panderichthys* is known from Latvia, where it lived some 385 million years ago (the Mid Devonian period).

Until very recently, the earliest tetrapod fossils that are more nearly fishlike were also from the Devonian, about 376 million years old. They have been found in Scotland and Latvia. *Ichthyostega* and *Acanthostega* from Greenland, which lived more recently, about 365 million years ago, are unambiguous tetrapods, with limbs that bear digits, although they retain from their fish ancestors such characteristics as true fish tails with fin rays. Thus, the time gap between the most tetrapod-like fish and the most fishlike tetrapods was nearly 10 million years, between 385 and 376 million years ago

The recently discovered *Tiktaalik* goes a long way towards breaching this gap; it is the most nearly intermediate between

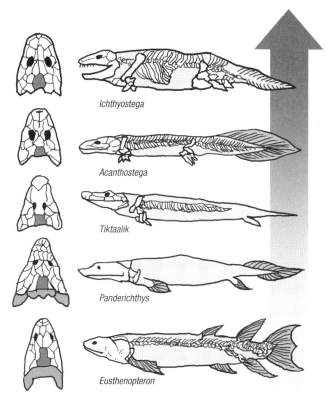

Ichthyostega

Acanthostega

Tiktaalik

Panderichthys

Eusthenopteron

TIKTAALIK AND OTHER FOSSIL INTERMEDIATES BETWEEN FISH AND TETRAPODS (ANIMALS WITH FOUR LIMBS), WHICH LIVED BETWEEN 385 (EUSTHENOPTERON) AND 359 (ICHTHYOSTEGA) MILLION YEARS AGO.

fishes and tetrapods yet known. Several specimens have been found in Late Devonian river sediments, dated about 380 million years ago, on Ellesmere Island in Nunavut, Arctic Canada. *Tiktaalik* is Inuit for 'big freshwater fish'. It displays an array of features that are just about as precisely intermediate between fish and tetrapods as one could imagine and exactly fits the time gap as well.

Extinction

The number of species living on Earth at present or at any time represents the balance between the species that originated in the past and those that became extinct. Many strange or bizarre animals of the past are revealed by the fossil record: dinosaurs, ammonites, trilobites and many others. Palaeontologists estimate that more than 99 per cent and perhaps more than 99.9 per cent of all species that lived in the past became extinct. That gives rise to a sobering thought. The number of species now living on Earth (excluding bacteria and archaea) is estimated at more than 10 million. If this number represents 1 per cent of the total that ever existed, that would amount to 1 billion species; if it only represents 0.1 per cent, the number of species that have lived on Earth would amount to 10 billion. Why did so many species become extinct?

> *The number of species living on Earth at present or at any time represents the balance between the species that originated in the past and those that became extinct.*

Natural selection acts to improve or maintain adaptation to prevailing conditions. If environmental change is very slow, then selection may be able to maintain a high level of adaptation. However, adaptations are not selected for future conditions, but for present ones. Environmental change is environmental deterioration so far as organisms are concerned. Since environmental change is incessant, we expect extinctions to occur continuously, rising in intensity during periods of greater and more rapid change.

Palaeontologists identify two patterns of extinction. One is the pattern of 'background extinctions', which reflects the failure of species to adapt to the ongoing process of environmental change, not only of the physical environment but also of the biotic environment, including competitors, parasites and predators. But there are also 'mass extinctions', when a very large number of taxa, even a majority in some cases, became extinct over a relatively short (in the geological scale) period of time.

Five mass animal extinctions are generally recognized; they occurred at the end of the Ordovician period, late in the Devonian, at the boundary between the Permian and the Triassic, at the end of the Triassic and at the boundary between the Cretaceous and the Tertiary (K/T extinction). The corresponding times for the five mass extinctions, in millions of years ago, are approximately 440, 370, 260, 200 and 65. Not all were equally severe, nor were all animal taxa equally impacted. The most severe episode of mass extinction was the one at the boundary between the Permian and the Triassic, some 260 million years ago, when trilobites became extinct, and corals, brachiopods and crinoids very nearly so. Estimates of the taxa extinguished around the boundary between the Permian and the Triassic are 54 per cent of marine families, 84 per cent of genera and 80–90 per cent of all species.

The causes of mass extinctions are for the most part conjectural. Massive volcanic eruptions throughout the planet are considered one likely cause of the event at the Permian/Triassic boundary, with associated global warming. Particularly interesting is the K/T extinction. It was suggested in 1980 that the cause was the impact of an extraterrestrial body, which would have extended a thick layer of dust throughout the atmosphere, darkening the sky and lowering temperatures. Large terrestrial animals, notably the dinosaurs, would have become extinct. The small mammals then in existence would have fared better and some survived. As the atmosphere recovered, the mammals diversified and larger taxa evolved, notably the primates and humans. Geologists have identified the Chicxulub crater off the coast of the Yucatan Peninsula of Mexico as the place where the impact occurred.

Think of it. The dinosaurs dominated the Earth, seemingly inhibiting the evolution of other large animals. According to some evolutionists, their extinction made possible the diversification in size and otherwise of the pre-existing small mammals. If so, we humans owe our existence to a meteorite. Its impact made it possible for primates and eventually humans to evolve.

WHAT IS THE MISSING LINK?

The story of human ancestors

*D*arwin extended the theory of evolution by natural selection *to humans in* The Descent of Man, *which he published in 1871, 12 years after* On the Origin of Species. *Fossils intermediate between humans and apes were yet to be discovered – the 'missing link' alleged by Darwin's critics. The missing link is no longer missing. Thousands of intermediate fossil remains (known as 'hominids') have been discovered since Darwin's time, and the rate of discovery is accelerating.*

Mankind is a biological species that has evolved from species that were not human. Our closest biological relatives are the great apes, primarily the chimpanzees, who are more closely related to us than they are to the gorillas, and much more than to the orang-utans. The hominid lineage diverged from the chimpanzee lineage 6–7 million years ago (Ma) and evolved exclusively in the African continent until the emergence of *Homo erectus* sometime before 1.8 Ma.

The missing link

Darwin's theory of evolution asserted that humans and apes share common ancestors that were not human. His contemporaries questioned the whereabouts of the 'missing link', the intermediate organism or organisms between apes and humans. Primates that were ancestors to humans after our lineage separated from the chimp lineage are called hominids (or hominins). At the time of Darwin's death in 1882, no hominid fossils ancestral to modern humans were known, although he was persuaded that they would eventually be found.

The first hominid fossil was discovered in 1889 by a Dutch physician, Eugene Dubois, on the island of Java. It consisted of a femur and a small cranium. Because he was expert in human anatomy, Dubois knew that these belonged to an individual with bipedal gait; the femur was very similar to the femur of a modern human. But the capacity of the small cranium was about 850–900 cc (cubic centimetres), which could hold a brain less than two pounds in weight, while the cranium of a modern human is about 1,300–1,400 cc (with a brain of about three pounds). The fossil discovered by Dubois was from an individual who lived about 1.8 Ma, and is now classified in the species *Homo erectus*. Our own species is called *Homo sapiens*.

The 'missing link' is no longer missing. The fossil from Java was the first one, but thousands of fossil remains belonging to thousands of individual hominids have been discovered in the twentieth and twenty-first centuries in Africa, Asia and Europe, and continue to be discovered. These fossils have been studied and dated, using radiometric and other methods. Some fossil hominids are very different from other fossils, as well as from humans, and are classified in separate species. The record of fossil hominids that lived at different times shows that several changes occurred through time in the lineage leading from non-human ancestors to modern humans. One change was increase in body size; another was bipedal gait; most important was increase in cranial capacity (and brain size). The species names are somewhat exotic, making reference sometimes to the place where the fossils were found, or to their morphological characteristics, or determined by the whim of the discoverers.

> *Our closest biological relatives are the great apes, primarily the chimpanzees, who are more closely related to us than they are to the gorillas.*

Ancestral hominids

The task of reconstructing the sequence of our hominid ancestors and distinguishing our ancestral lineage from hominids that were lateral lineages that became extinct is daunting. Although thousands of fossil hominids have been discovered, they are fragmentary and their geographic distinction and temporal sequence often ambiguous. As more and more fossils are discovered and described, the emerging picture of hominid evolution tends to be modified, although some lines of descent and lateral relationships are accepted by most experts. In addition to establishing lineage sequences, anthropologists seek to determine the evolution of features as they become (or not, in lateral lineages) more and more human-like. At stake are such issues as mode of locomotion, arm/hand and brain development and other anatomical features, as well as questions of diet, family organization, geographic distribution and many others.

> *The record of fossil hominids that lived at different times shows that several changes occurred through time in the lineage leading from non-human ancestors to modern humans.*

The oldest known fossil hominids are *Orrorin tugenensis* and *Sahelanthropus tchadensis*, both first described early in the twenty-first century (2001 and 2002). They lived about six to seven million years ago, and their anatomy indicates that they were predominantly bipedal when on the ground, but had very small brains, perhaps about the size of a chimpanzee's. Only 12 specimens of *O. tugenensis* are known, including several teeth, mandible fragments and arm and leg bones. The known remains of *S. tchadensis* are also precarious: nine specimens, which include a cranium, two mandibular fragments and several teeth.

Ardipithecus kadabba and *Ardipithecus ramidus* lived more recently, between 4.5 and nearly 6 million years ago. They are represented by several score fossils, including an almost complete skeleton of *A. ramidus*, described in 2009 and known as ARA-

VP-6/500, after Aramis, the African locality where it was found. The fossils indicate a behavioural mixture of arboreal climbing with bipedal walking when on the ground.

Numerous fossil remains from diverse African origins exist of *Australopithecus*, a hominid that appeared about 4 million years ago. *Australopithecus* had an upright human stance but a cranial capacity of about one pound, comparable to that of a gorilla or chimpanzee and about one third that of modern humans. The skull displayed a mixture of ape and human characteristics – a low forehead and a long, apelike face but with teeth proportioned like those of humans. Other early hominids partly contemporaneous with *Australopithecus* include *Kenyanthropus* and *Paranthropus*; both had comparatively small brains, although some species of *Paranthropus* had larger bodies. *Kenyanthropus* and *Paranthropus* represent side branches of the hominid lineage that became extinct.

Lucy

In 1976 the world was introduced to Lucy, the whimsical name given to the fossil remains of remains of a hominid ancestor classified as *Australopithecus afarensis*, a species of bipedal hominid, small-brained and some 3.5 feet tall. Lucy is famous because about 40 per cent of the skeleton of this young woman was found on a single site when it was discovered 35 years ago. Experts generally agree that *A. afarensis*, who lived between 3 and 3.6 million years ago, is in the line of descent to modern humans.

Australopithecus africanus, which lived more recently than *A. afarensis* and is the first *Australopithecus* species ever discovered, was also short and small-brained. However, *A. africanus* is not our ancestor, but is rather a co-lateral relative, the likely ancestor of *Australopithecus (Paranthropus) robustus* and other co-lateral hominids, who lived for 2 million years or more after their divergence from our ancestral lineage, and thus long coexisted in Africa with some of our ancestors (*A. afarensis, H. habilis* and *H. erectus*; see below). Some of these co-lateral relatives became somewhat taller and more robust, but their brains remained small, about 500–600 cc (less than 1.5 pounds) at the most.

Australopithecus anamensis, dated 3.9–4.2 million years ago, is commonly assumed to be the ancestral species to *A. afarensis*, whose earliest definitive specimen is dated as ~3.6 million years old. The analysis and publication on 13 April 2006 of 30 hominid specimens, representing a minimum of eight individuals, of *A. anamensis* from the Afar region of Ethiopia, dated to ~4.12 million years ago, supports the interpretation that *A. anamensis* is the ancestral species of *A. afarensis*. Moreover, these new fossils suggest that *Ardipithecus* was the most likely ancestor of *A. anamensis* and all later australopithecines. The fossils suggest that a relatively rapid evolution occurred from *Ardipithecus* to *Australopithecus* in this region of Ethiopia.

Much more similar to us are hominids classified as *Homo habilis*, the earliest species classified in the same genus as our own. *Homo habilis* individuals made very simple stone tools, the first hominids to do so, which is why they were given the name *habilis*, Latin for 'handy' or 'skilled'. Early *H. habilis* had a cranial capacity of about 600 cc, or somewhat larger, greater than any of the earlier hominids, but about half or less than half the brain size of modern humans. *Homo habilis* lived in tropical Africa between 2.5 and 1.5 Ma. In *Homo habilis* we can see the trifling beginnings of human technology.

The discovery of *H. habilis* early in the 1960s changed the understanding of hominid evolution. *H. habilis* exhibits features that represent changes in the cranium and dentition compared with *Australopithecus*. Its face is less prognathic and its cranial capacity is larger. Its masticatory apparatus is smaller, especially molars and premolars, and the dental enamel is slightly thinner. The shape of its dental arcade is parabolic, like in later *Homo* specimens. Some palaeontologists thought that the *H. habilis* specimens should be considered *Australopithecus*. Others thought they should be classified as *H. erectus*.

These contrasting views about *H. habilis* were, in a way, the best argument in favour of the proposal. The new specimens exhibited an intermediate morphology, and their inclusion in one or other side depended on the emphasis placed on similarities and differences. The new specimens could not unequivocally be considered either *Australopithecus* or *H. erectus*. A new species was required, but why in the genus *Homo*?

The genus has become associated with features other than morphological traits, namely, the production of tools used for scavenging and hunting. This behaviour requires a big enough brain to carry out the complex cognitive operations involved in such tasks. The proponents of the new taxon suggested that *H. habilis* was the true author of the Oldowan culture, the lithic industry at Olduvai. *Homo* would be the genus that introduced the adaptive strategy of stone tool-making, and *H. habilis* its first representative.

Homo habilis was succeeded by *Homo erectus*, which evolved in Africa sometime before 1.8 Ma, had a cranial capacity of 800–1,100 cc (2–2.5 pounds) and made tools more advanced than those of *Homo habilis*. Two features of *Homo erectus* deserve particular attention. One is that the species persisted (with relatively small morphological changes at various times and places) for a long time, from 1.8 Ma up to nearly 400,000 years ago. A second distinctive feature of *Homo erectus* hominids is that they were the first intercontinental wanderers among our hominid ancestors. Shortly after their emergence in Africa, they spread to Europe and Asia, reaching as far as northern China and Indonesia (Java, where Eugene Dubois found the first hominid fossils ever discovered). *Homo erectus* fossils from Indonesia have been dated at 1.81 and 1.66 Ma and from Georgia (in Europe, near the Asian frontier) between 1.6 and 1.8 Ma.

Several species of hominids lived in Africa, Europe and Asia between 1.8 million and 500,000 years ago. They are known as *Homo ergaster*, *Homo antecessor* and *Homo heidelbergensis*, with brain sizes roughly that of the brain of *Homo erectus*. Some of these species were partly contemporaneous, though they lived in different regions of the Old World. These species are sometimes included under the name *Homo erectus sensu lato* (meaning 'in a broad sense').

Modern humans

Two hominid species that evolved after *H. erectus* are *Homo neanderthalensis* and *Homo sapiens*, our own. Numerous *Homo neanderthalensis* ('Neanderthal man') fossils are known from Europe, where they first appear about 200,000 years ago, becoming extinct 30,000 years ago. The most recent fossils of *Homo neanderthalensis* were found in Spain, where they seemingly had their last abode. Neanderthals had large brains, much like ours, and bodies also similar to ours, but somewhat stockier.

The evolution from *Homo erectus* to *Homo sapiens* may have started about 400,000 years ago, when fossils are found that are considered 'archaic' forms of *Homo sapiens*. Anatomically modern humans evolved in Africa around 200,000 or 150,000 years ago and eventually colonized the rest of the world, replacing other hominids. The *Homo erectus* who had earlier colonized Asia and Europe did not leave any direct descendants. (A possible exception is *Homo floresiensis*, the minuscule hominids whose fossil remains were discovered in 2004 on the Indonesian island of Flores, where they lived 12,000–18,000 years ago. They may have been direct descendants of Asiatic *Homo erectus*, although the matter is still being investigated.)

The Neanderthals have been thought to be ancestral to anatomically modern humans, but we now know that modern humans appeared more than 100,000 years ago, long before the disappearance of Neanderthal fossils. It is puzzling that, in caves in the Middle East, fossils of anatomically modern humans precede as well as follow Neanderthal fossils. Some modern humans from these caves are dated at 120,000–100,000 years ago, whereas Neanderthals are dated at 60,000 and 70,000 years, followed by modern humans dated at 40,000 years. It is unclear whether Neanderthals and modern humans repeatedly replaced one another by migration from other regions, or whether they coexisted, and indeed whether interbreeding may have occurred (although comparisons of DNA from Neanderthal fossils with living humans indicate that no, or very little, interbreeding occurred between Neanderthals and their contemporary, anatomically modern humans).

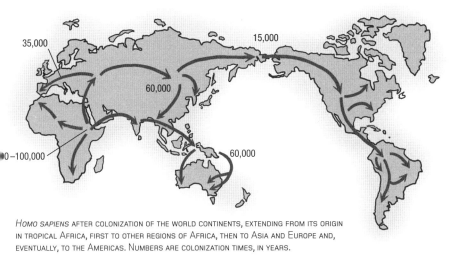

35,000

15,000

60,000

0–100,000

60,000

Homo sapiens after colonization of the world continents, extending from its origin in tropical Africa, first to other regions of Africa, then to Asia and Europe and, eventually, to the Americas. Numbers are colonization times, in years.

Some anthropologists have argued that the transition from *H. erectus* to archaic *H. sapiens*, and later to anatomically modern humans, occurred concurrently in various parts of the Old World (Africa, Asia and perhaps Europe). Most scientists believe instead that modern humans first arose in Africa somewhat earlier than 100,000 years ago and spread from there throughout the world, replacing the pre-existing populations of *H. erectus* and related hominid species, including, later, *H. neanderthalensis*. Some proponents of this African replacement model argue further that the transition from archaic to modern *H. sapiens* was associated with a reduction of the human population to a relatively small number of individuals, and that these small numbers of individuals are the ancestors of all modern humans (see *Am I Really a Monkey?*).

Recently, analyses of DNA from living humans have confirmed the African origin of modern *H. sapiens*, dating it at about 156,000 years ago in tropical Africa. Shortly thereafter, modern humans spread through Africa and throughout the world. South East Asia and the region that is now China were colonized by 60,000 years ago. Shortly thereafter, modern humans reached Australasia. Europe was colonized more recently, only about 35,000 years ago, and America even more recently, perhaps only 15,000 years ago. Ethnic differentiation between modern human

137

populations is therefore evolutionarily recent, a result of divergent evolution between geographically separated populations during the past 50,000 to 100,000 years.

Human genetic diversity

One hundred thousand years encompasses about 5,000–6,000 hominid generations, which is not a long time on the evolutionary scale. Thus, if the dispersal of modern humans from Africa to the rest of the world began 100,000 years ago, we would expect that the genetic differentiation among human populations should not be very large, even if we exclude intermingling between populations, which is occurring at an increasing rate in modern times.

Scientists have discovered that the genetic diversity among human populations in different parts of the world is only about 15 per cent higher than among people from the same village. This might at first seem surprising, because we are aware of the conspicuously different appearance of humans from different regions of the world (the human races or ethnic groups), but it is less unexpected when we take into account the fact that the divergence of human populations is of recent origin.

Ethnic differentiation between modern human populations is therefore evolutionarily recent, a result of divergent evolution between geographically separated populations during the past 50,000 to 100,000 years.

Of the total genetic variation of all of humankind, 85 per cent is present among individuals of the same population, say, of the same village or town. (This is without taking into account interbreeding with migrants from other populations, which does augment the percentage above 85.) Approximately 6 per cent additional variation is found among people from different localities on the same continent, and an additional 9 per cent of the variation is found among individuals from different continents.

As I have pointed out, these results may be expected because of the evolutionarily recent dispersal of human populations, but they seem to contradict common experience. We know that tropical Africans look quite different from Scandinavians, and both look very different from Japanese people. The explanation of this conundrum has two components. First, our African ancestors were already genetically quite varied by the time they began to colonize the rest of the world. This is not unexpected, because such is the case with most animal species: genetically they vary a great deal. Indeed, chimpanzees are genetically more varied than humans, although the total world population of chimpanzees is much smaller than the seven billion humans. Ancestral *H. sapiens* populations retained most of their original genetic variation as they colonized different regions of the world. Second, the stereotype traits, such as skin colour, hair colour and texture and facial features, that distinguish ethnic groups involve relatively few genes. Some of these genes have evolved as adaptations in response to different climates. Consider, for example, one of the most conspicuous differences among ethnic groups: skin pigmentation.

Melanomas are severe cancers caused by sustained exposure to the sun's ultraviolet radiation. Thus, peoples living for generations at low latitudes have genes that produce greater amounts of eumelanins (brown and black melanin), which filter out most UV radiation and thus protect the skin from damage. On the other hand, some UV radiation is necessary for the synthesis of vitamin D in the deeper layers of the skin. Thus, the amount of eumelanin that is adaptive in the tropics is less than optimal at high latitudes, where UV radiation is much lower. At high latitudes, natural selection has favoured genes that result in pale skin, so that UV reaches the layers of the dermis where vitamin D is synthesized. Examples like this one have helped demolish the myth of great genetic differentiation between 'races'. It is just that: a myth without scientific support.

IS INTELLIGENCE INHERITED?

Individual intelligence depends only partly on genes

If a person who is not a scientist is asked to choose what they think is the most 'human' trait, the one that most distinguishes humans from apes, that person would probably choose some feature related to the mind. It could be the size of the brain, which is about four times larger in humans than in chimpanzees or gorillas, or it could be some functional trait, like language, complex social organization or technology. The person might perhaps say 'exalted intelligence', which depends on our large brain and is related to the other functional features.

We know that eye colour, skin pigmentation, blood group, body size and many other human features are inherited from our parents to a greater or lesser extent. But what about behavioural features and, in particular, intelligence? Do we inherit intelligence from our parents, or does it develop mostly as a consequence of schooling, learning and other human experience? And can intelligence be measured? How?

Heredity and environment

In order to understand biological heredity, it is necessary to make the important distinction between genotype and phenotype. The phenotype of an organism is its appearance – what we can observe: its morphology, physiology and behaviour. The genotype is the genetic constitution it has inherited. During the lifetime of an individual, the phenotype may change; the genotype, however, remains constant. The genotype is made up of genes, the DNA sequences that we have inherited from our parents.

The distinction between phenotype and genotype must be kept in mind, since the relation between the two is not fixed. This is because the phenotype results from complex networks of interactions between different genes, and between the genes and the environment, and the outcome will be different in different environments. Human individuals do not have identical phenotypes, although the phenotypes may be similar when only one or a few traits are considered. For example, two people may both have brown eyes or group A blood (haemoglobin). Organisms having similar phenotypes with respect to a given trait do not necessarily have identical genotypes even with respect to that trait. For example, people with brown eyes may have two genes for brown eye

The phenotype of an organism is its appearance – what we can observe: its morphology, physiology and behaviour. The genotype is the genetic constitution it has inherited.

colour or may have inherited the gene for brown eye colour from one parent and the gene for blue eye colour from the other. The gene for brown eye colour is dominant over the gene for blue eye colour, so that only the gene for brown colour is expressed in the individual. Similarly, a person with group A blood may have two genes for A or may have one gene for A and the other for group O. The gene A is dominant over the O gene. The genes for blue eye and for O blood group, although they are not expressed in individuals with two different genes, may nevertheless be inherited by their children. Each of the two genes, brown and blue or group A and group O, has equal probability of being passed to their children.

Some human hereditary traits are discrete, such as eye colour or blood group. There are no persons who have all sorts of intermediate eye colours ranging from brown to blue. People may have blood group A, B or O, but there are no persons with blood

THE NORMAL DISTRIBUTION. HEIGHT DISTRIBUTION OF A GROUP OF MALE (BLACK SWEATER) AND FEMALE (WHITE SWEATER) COLLEGE STUDENTS. MOST STUDENTS ARE OF INTERMEDIATE HEIGHT, WITH FEWER AND FEWER TOWARD THE EXTREMES.

groups intermediate between any two of them. However, not all traits appear in clearly distinguishable alternative forms. Humans do not come in only two height classes, tall and short; rather, they vary continuously over a large range of heights. Height, weight, fertility and longevity are examples of many traits that exhibit more or less continuous variation so that typically a majority of individuals have intermediate values and there are fewer and fewer individuals exhibiting more and more extreme values, say either very, very tall or very, very short. This common pattern of variation is known as the 'normal distribution'. Intelligence is clearly a trait with continuous variation between very intelligent people and those who have very low intelligence.

The occurrence of continuous variation may be due to interactions between genes and the environment, or interactions between different genes. A person who eats more and exercises less is likely to be heavier than a person who eats less and exercises more. People with university degrees display higher intelligence than people who did not go to school or receive much formal education from their parents. But we also need to take into account that there are various genes that impact weight, and there may be a variety of genes that impact intelligence. Given the important role that education plays in determining intelligence, how do we ascertain whether heredity plays a role, and how do we assess the relative importance of genes relative to education and experience in determining intelligence? Is it possible to ascertain to what extent the intelligence of an individual is determined by the genes and to what extent by the environment? But before we answer that question, we need to know whether one can measure intelligence and how. What is intelligence, anyway?

Intelligence quotient

There is no definition of intelligence that would be generally accepted by experts, nor is it likely that anybody would have a clear and consistent idea of what it is that we mean by intelligence. Intelligence refers to cognitive ability and is often seen as a unitary capacity, but it encompasses a multiplicity of abilities that are usually not developed to the same extent as each other. A person with well-developed linguistic abilities may have limited mathematical capacity, and a person with sophisticated analytical powers may be poor at intuition or creativity. The theorist J.P. Guilford carried this multidimensional approach to an extreme by identifying 120 highly specific abilities that jointly defined intelligence. These abilities included remembering, divergent thinking, processing information concerning, for example, letters, words and numbers, as well as the products of processing the information, such as relations, classes and systems.

Intelligence tests have been developed over the last century in an effort to achieve some objective measure of intelligence. The most commonly used tests seek to determine a person's intelligence quotient, IQ. The IQ is the percentage obtained by dividing a person's mental age by their chronological age. The mental age is determined by testing a set of cognition-related skills and then comparing the score obtained by an individual to the average obtained by a number of people of the same age. The average score is set at 100. A higher or lower score indicates higher or lower than average intelligence. A person with a score of 130 or above is considered gifted and a person with a score of 70 or lower is considered mentally deficient. Other tests

Intelligence refers to cognitive ability and is often seen as a unitary capacity, but it encompasses a multiplicity of abilities that are usually not developed to the same extent as each other.

have been developed, but there is no agreement as to which one, if any, may provide a satisfactory measure of intelligence.

Heritability

We have noted earlier that both heredity and environment contribute to determining the phenotype of a person. But the relative importance of the genes and the environment may vary considerably from trait to trait. Traits largely determined by single genes are usually little impacted by the life experience of the individual, which starts at conception and extends through pregnancy, birth and the full life until death. Whether a person has brown or blue eyes depends overwhelmingly on the genes inherited from the parents. Similarly, a person's life experience is of little consequence in determining their blood group; what counts is the genotype, the genes inherited.

The expression of other traits, however, depends on the environment to a greater or lesser extent. These are typically traits whose expression is, in turn, impacted by several or many genes, called quantitative genes, or polygenes (meaning many genes). The traits themselves are quantitative traits, because their expression varies gradually from individual to individual. Examples already mentioned are a person's weight and height, but also intelligence. A well-schooled person will have a higher IQ than one raised in a deprived environment. But at issue is whether IQ might also depend on the genes that a person inherited.

Is it possible to ascertain to what extent intelligence (or any quantitative trait) is determined by the genotype and to what extent by the environment? Reflecting on this question makes us realize that it is not well formulated. Any trait depends completely for its development on both heredity *and* environment. For an individual to develop, it must have a genotype – the genetic constitution inherited from the parents at conception – but development can only occur in interaction with the environment. There is no way in which we could measure the IQ of a human genotype that has not been exposed to any environment whatsoever; such individuals cannot exist.

The question of the relative effects of heredity and environment is sometimes raised as the question of 'nature versus nurture'. It can be properly expressed as follows: to what extent is the variation among individuals with respect to a trait due to genetic variation (i.e. genetic differences among individuals), and to what extent is it due to environmental variation (i.e. environmental differences)? It will soon be apparent why it is important to realize that this question about variation among individuals, rather than the one formulated earlier about the intelligence of an individual, is the one being answered when geneticists investigate the relative effects of heredity and environment.

The fraction of the phenotypic variation in a trait that is due to genetic differences can be measured by the heritability of the trait, a concept first advanced by geneticists working on plant and animal breeding who asked, for example, to what extent the number of eggs laid by a hen or the amount of milk produced by a cow would depend on the breed, the genetic make-up, and to what extent it would depend on how they are fed, housed and otherwise treated on the farm.

We'll now introduce symbols that will help us to deal with the heritability issue.

H = heritability

V_T = the total phenotypic variance observed in a trait

V_G = the fraction of the phenotypic variance that is due to genetic differences among individuals

V_E = the fraction of the phenotypic variance that is due to differences in the environmental conditions to which the individuals were exposed

We thus have $V_T = V_G + V_E$; that is, the total variance of the trait is simply the variance contributed by the genes plus the variance contributed by the environment.

By definition:

$$\text{Heritability} = \frac{\text{Genetic variance}}{\text{Phenotypic variance}} = \frac{V_G}{V_T} = \frac{V_G}{V_G + V_T}$$

That is precisely what we want to find out: what fraction of the total variance is due to genetic differences among individuals.

'Variance' is a term used by statisticians because it is useful for summarizing and understanding quantitative variation in many situations, particularly in organisms. It is fairly easy to calculate. First, one needs to calculate the 'mean', which is the average value of the trait under consideration. If we have measured the IQ in, say, 60 people, we add all the IQ values and divide them by 60 to obtain the mean or average IQ in that group of people; say we obtain a mean IQ value of 103. The variation among individuals is measured by the variance. But here we find a small trick, invented by statisticians for good reasons, which we don't need to justify here. The difference between each individual and the mean is first obtained and then squared. All squared differences are added up and divided by the number of individuals, 60 in the example. The average of the squared differences is the variance of the trait among the 60 individuals in that group.

> *For an individual to develop, it must have a genotype – the genetic constitution inherited from the parents at conception – but development can only occur in interaction with the environment.*

Obtaining the mean and variance of a group of individuals (a 'population') is simple and straightforward as just described. But determining how much of the variation is due to the environment and how much is due to the genetic differences among individuals is not a simple matter. One method used by geneticists for this purpose is called the 'twin method'.

The twin method

One way of estimating heritability is to measure the phenotypic variance in groups of relatives with known degrees of relatedness, such as twins or first cousins. There are two kinds of twins. Identical twins arise from a single fertilized ovum that early on splits into two embryos that develop independently, resulting in two genetically identical individuals. Fraternal twins arise from two independent zygotes, i.e. two eggs fertilized by two different sperm.

The genetic relationship between fraternal twins is the same as that between ordinary full-sibs (siblings), because in both instances zygotes form from different eggs from the same mother and different sperm from the same father. Fraternal twins, like full-sibs, share on average half their genes (for each gene received by one sib from the mother, the probability that the other sib will receive the same gene is one half, and the same for the paternal genes).

Identical twins are always of the same sex; fraternal twins may be both female, one female and one male, or both male, in the proportions 1:2:1, the same as for brothers and sisters who are not twins. In order to estimate heritability, sets of identical twins of one sex are compared with sets of fraternal twins of the same sex as the identical twins. Let V_I = phenotypic variance between identical twins and V_F = phenotypic variance between fraternal twins. Because the identical twins are genetically identical, their variance is all environmental: $V_I = V_E$.

The variance between fraternal twins is part genetic and part environmental. However, because they have half of their genes in common, their genetic variance will be half that of unrelated individuals:

$$V_F = V_G/2 + V_E$$

Therefore, the difference between the two phenotypic variances estimates half the genetic variance:

$$V_F - V_I = V_G/2 + V_E - V_E = V_G/2$$

Twice that difference divided by the total phenotypic variance estimates heritability:

$$H = \frac{2(V_F - V_I)}{V_T} = \frac{2(V_G/2)}{V_T} = \frac{V_G}{V_T}$$

The procedure seems straightforward, but there are, nevertheless, problems. The environmental variance has been assumed to be the same for identical and for fraternal twins of the same sex, because in both cases the twins are born at the same time, are of the same sex, live together, go to the same schools, etc. But it is possible that identical twins are treated more similarly than fraternal twins, precisely because identical twins are genetically more alike and thus tend to react more similarly. The formula given above will then overestimate heritability.

Another difficulty is that twins are also treated more similarly than unrelated individuals because the twins are brought up in the same family, cultural milieu, etc. This problem can be partially circumvented by studying twins adopted by different families. Although the proportion of such cases is limited, a number of them have been studied by geneticists.

Estimates of IQ heritability obtained in different studies vary, but usually within a relatively small range around 0.70. To the extent that we are willing to accept that IQ is a worthy measure of intelligence, we may conclude that intelligence is strongly inherited, meaning, more precisely, that the variation in IQ that we observe among individuals is, in good part, due to differences between the genes they inherited. Within any given human population, it seems that 70 per cent of the variation in intelligence is due to genetic factors and only 30 per cent due to environmental factors. Keep in mind that these estimates refer to individuals of similar age and fairly similar education and socioeconomic status, so that the variation due to the environment is similar among them.

We may place IQ heritability estimates in context by comparing them with heritability estimates for other human traits. The heritability of stature is estimated around 0.85, among the highest for any human quantitative trait. The heritability of body weight is estimated at around 0.62, of systolic blood pressure at 0.57 and of diastolic blood pressure at 0.63.

A wider context can be provided by heritability estimates for various traits in various organisms: slaughter weight in cattle raised on the same ranch, 0.85; plant height in corn, 0.70; egg weight in poultry, 0.60; fleece weight in sheep, 0.40; milk production in cattle, 0.30; yield in corn, 0.25; ear length in corn, 0.17. Heritability estimates are quite useful in plant and animal breeding because they indicate the amount of response that farmers can expect from selective breeding of desirable traits; that is by breeding the cows that produce the most milk, or by planting corn seeds from plants with higher yields.

In conclusion, it bears repetition that heritability is a population-specific measurement. The high heritability of IQ tells us that in a given population, much of the variation in IQ values is due to genetic differences among individuals, and we may therefore infer that intelligence is to a certain extent inherited. But it does *not* tell us that intelligence is 70 per cent hereditary and 30 per cent environmental. The point is obvious if we focus on two populations with disparate socioeconomic status, education, health care and the like. Think of human populations in poor countries, where most individuals are undernourished, receive virtually no health care and little if any schooling, suffer from severe endemic diseases and have considerable infant mortality. It would be grossly misplaced to assert that the difference in IQ or other expressions of intelligence between individuals there and those in a developed country is 70 per cent due to genetic differences. It might be much more nearly correct to assert that intelligence differences among individuals in the poor country are largely due to hereditary differences, as they are also in the developed country.

WILL HUMANS CONTINUE TO EVOLVE?
Biological and cultural evolution

*H*umans are animals, but a very distinct and unique kind of
animal. Our anatomical differences include bipedal gait and
enormous brains. We are notably different also, and more importantly,
in our individual and social behaviour. Biological evolution persists
in modern humans, but the advent of humankind ushered in cultural
evolution, a more rapid and effective method of adaptation than
the biological mode. Products of cultural evolution include science
and technology; social and political institutions; religion and ethics;
language, literature and art; and electronic communication.

Humans are members of the species *Homo sapiens*, which evolved
from species that were not human. The evolution of the hominid
lineage, from our ancestors of 6–7 million years ago to the present,
has been considerable by any standards. It has entailed numerous
and important anatomical changes, which in turn account for
radical behavioural changes.

Natural selection in modern humans

There is no scientific basis to the claim sometimes made that
the biological evolution of mankind has stopped, or nearly so,
at least in technologically advanced countries. It is asserted that
the progress of medicine, hygiene and nutrition has largely
eliminated death before middle age; that is, most people live
beyond reproductive age, after which death is inconsequential for

natural selection. That mankind continues to evolve biologically can be shown because the necessary and sufficient conditions for biological evolution persist. These conditions are genetic variability and differential reproduction. There is a wealth of genetic variation in mankind. With the exception of identical twins, developed from a single fertilized egg, no two people who live now, lived in the past or will live in the future are likely to be genetically identical. Much of this variation is relevant to natural selection.

Does natural selection continue to occur in modern mankind? Natural selection is simply differential reproduction of alternative genetic variants. Therefore, it will occur in mankind if the carriers of some genotypes are likely to leave more descendants than the carriers of other genotypes. Natural selection consists of two main components: differential mortality and differential fertility; both persist in modern mankind, although the intensity of selection due to postnatal mortality has been somewhat attenuated.

Death may occur between conception and birth (prenatal) or after birth (postnatal). Death during the early weeks of embryonic development may go totally undetected. But it is known that about 20 per cent of all ascertained human conceptions end in spontaneous abortion during the first two months of pregnancy. Such deaths are often due to deleterious genetic constitutions, and thus they have a selective effect in the population. The intensity of this form of selection has not changed substantially in modern mankind, although it has been slightly reduced with respect to a few genes, such as those involved in Rh blood group incompatibility.

Postnatal mortality has been considerably reduced in recent times in technologically advanced countries. For example, in the United States, somewhat less than 50 per cent of those born in 1840 survived to age 45, while the average life expectancy for people born in 1960 is 78 years. In other regions of the world, postnatal mortality remains quite high, although it has generally decreased in recent decades. Mortality before the end of reproductive age, particularly where it has been considerably reduced, is largely associated with genetic defects, and thus it has a favourable selective effect in human populations. More than 4,000 genetic variants are known that cause diseases and malformations in humans; such variants are kept at low frequencies due to natural selection.

It might seem at first that selection due to differential fertility has been considerably reduced in industrial countries as a consequence of the drop in the average number of children per family that has taken place. However, this is not so. The intensity of fertility selection depends not on the mean number of children per family, but on the *variance* in the number of children per family. It is clear why this should be so. Assume that all people of reproductive age marry and that all have exactly the same number of children. In this case, there would not be fertility selection independent of whether couples all had very few or very many children. Assume, on the other hand, that the mean number of children per family is low, but some families have no children at all while others have many. In this case, there would be considerable opportunity for selection – the genotypes of parents producing many children would increase in frequency at the expense of those having few or none. Studies of human populations have demonstrated that the opportunity for natural selection often increases as the mean number of children decreases. An extensive study published in 1961 showed that the index of opportunity for selection due to fertility was four times larger among United States women born in the twentieth century, with an average of less than three children per woman, than among women in the Gold Coast of Africa or in rural Quebec, who on average had seven children. There is no evidence that natural selection due to fertility has decreased in modern human populations.

Natural selection may decrease in intensity in the future, but it will not disappear altogether. So long as there is genetic variation and the carriers of some genotypes are more likely to reproduce than others, natural selection will continue. Cultural changes, such as the development of agriculture, migration from the country to the cities, environmental pollution and many others, create new selective pressures. The stresses of city life are partly responsible for the high incidence of mental disorders in certain human societies. The point to bear in mind is that human environments are changing faster than ever, owing precisely to the accelerating rate of cultural change; and environmental changes create new selective pressures, thus fuelling biological evolution.

> *So long as there is genetic variation and the carriers of some genotypes are more likely to reproduce than others, natural selection will continue.*

Hereditary variation

As we know, natural selection is the process of differential reproduction of alternative genetic variants. In terms of single genes, variation occurs when two or more alleles are present in the population at a given gene locus. (Alleles are the variant forms of a particular gene.) How much genetic variation exists in the current human population? The answer is 'quite a lot', as will be presently shown, but natural selection will take place only if the alleles of a particular gene have different effects on fitness; that is, if alternative alleles differentially impact the probability of survival and reproduction.

The two genomes that we inherit from each parent are estimated to differ at about 1 or 2 nucleotides per thousand. The human genome consists of somewhat more than 3 billion nucleotides. Thus, about 3–6 million nucleotides are different between the two genomes of each human individual, which is a lot of genetic polymorphism. Moreover, the process of mutation introduces new variation in any population every generation. The rate of mutation in the human genome is estimated to be about 10^{-8}, one nucleotide mutation for every 100 million nucleotides, or about 30 new mutations per genome per generation. Thus, every human has about 60 new mutations (30 in each genome) that were not present in their parents. If we consider the total human population, that is 60 mutations per person multiplied by 7 billion people, or about 420 billion new mutations per generation, which are added to the pre-existing 3–6 million polymorphic nucleotides per individual, amounting to about 21–42 million billion polymorphic nucleotides among the human beings living in the world today.

That is a lot of mutations, but we must remember that the polymorphisms that count for natural selection are those that impact the probability of survival and reproduction of their carriers. Otherwise, the variant nucleotides may increase or decrease in frequency by chance, a process that evolutionists call 'genetic drift,' but will not be impacted by natural selection.

Genetic disorders

More than 4,000 human diseases and abnormalities that have a genetic causation have been identified in the human population. Genetic disorders may be dominant, recessive, multifactorial or chromosomal. Dominant disorders are caused by the presence of a single copy of the defective allele, so that the disorder is expressed in heterozygous individuals, those having one normal and one defective allele. In recessive disorders, the defective allele must be present in both alleles; that is, it is inherited from each parent. Multifactorial disorders are caused by interaction among several gene loci. And chromosomal disorders are due to the presence or absence of a full chromosome or a fragment of a chromosome.

Examples of dominant disorders are some forms of retinoblastoma and other kinds of blindness, achondroplastic dwarfism and Marfan syndrome (which is thought to have affected President Lincoln). Examples of recessive disorders are cystic fibrosis, Tay-Sachs disease and sickle-cell anaemia (caused by an allele that in heterozygous conditions protects against malaria). Examples of multifactorial diseases are spina bifida and cleft palate. Among the most common chromosomal disorders are Down's syndrome, caused by the presence of an extra chromosome 21, and various kinds of disorders due to the absence of one sex chromosome or the presence of an extra one, beyond the normal condition of XX for women and XY for men. Examples are Turner's syndrome (XO) and Klinefelter's syndrome (XXY).

The incidence of genetic disorders in the living human population is estimated to be no less than 2.56 per cent, impacting about 180 million people. Natural selection reduces the incidence of the genes causing disease, more effectively in the case of dominant disorders, where all carriers of the gene will express the disease, than for recessive disorders, which are expressed only in homozygous individuals. Consider, for example, phenylketonuria, a lethal disease if untreated, which has an incidence of 1 in 10,000 newborns, or 0.01 per cent. The frequency of the allele is about 1 per cent, so that in heterozygous conditions it is present in more than 100 million people, but only the 0.01 per cent of people who are homozygous express the disease and are subject to natural selection. The reduction of genetic disorders due to natural selection is balanced by their increase due to the incidence of new mutations.

Beyond disease

Natural selection acts also on a multitude of genes that do not cause disease. Genes impact skin pigmentation, hair colour and configuration, height, muscle strength and body shape and many other anatomical polymorphisms that are apparent, as well as many that are not externally obvious, such as variations in blood group, in the immune system and in the heart, liver, kidney, pancreas and other organs. It is not always known how natural selection impacts these traits, but we know that it does so differently in different parts of the world or at different times, as a consequence of the development of new vaccines, drugs and medical treatments; and also as a consequence of changes in lifestyle, such as the reduction in the number of smokers or the increase in the rate of obesity in a particular country.

There are about 20,000–30,000 human genes that encode enzymes and other proteins. Each protein may consist of about 300 amino acids, on average, which amounts to 900 nucleotides per gene. (Each amino acid is encoded by a 'triplet', three consecutive nucleotides.) Some nucleotide mutations do not change the encoded amino acids (because the 'genetic code' is redundant, so that two or more different triplets may code for the same amino acid). But about two thirds of all nucleotide mutations will change the encoded amino acid, and a majority of these will be subject to natural selection, either in small ways or substantially as in the case of genetic diseases. Consider now that that rate of mutations per nucleotide is 10^{-8} (one per 100 million nucleotides). We estimated that the number of nucleotides coding for each enzyme or protein is about 900, or 18–27 million nucleotides for all 20,000–30,000 genes encoding proteins, of which only two thirds, or about 12–18 million nucleotides, change the encoded amino acids. If the rate of mutation per nucleotide is 10^{-8}, the probability that a human will have a new amino-acid-changing mutation is about 24–36 per cent (because we have two copies of each gene). The nearly 7 billion humans now living will collectively have about 2 billion new amino-acid-changing mutations. These mutations, added to the pre-existing polymorphisms, provide natural selection with virtually unlimited opportunities. There can be little doubt that biological evolution continues to occur in humankind.

No species before mankind could select its own evolutionary destiny; mankind possesses techniques to do so, and more powerful techniques for directed genetic change are becoming available.

Where is human evolution going? Biological evolution is directed by natural selection, which is not a benevolent force guiding it towards sure success. The end result may be extinction. More than 99.9 per cent of all species that existed since the origin of life on Earth have become extinct. Natural selection has no purpose; humans alone have purpose and humans alone can introduce this purpose into their evolution. No species before mankind could select its own evolutionary destiny; mankind possesses techniques to do so, and more powerful techniques for directed genetic change are becoming available. Because we are self-aware, humans cannot refrain from asking what lies ahead, and because we are ethical beings, we must choose between alternative courses of action, some of which may appear as good, others as bad. The advances in genetic knowledge, molecular biology, medicine and associated techniques will surely be used in the future much more extensively and aggressively than they are now. It has been suggested that such advances could be used to 'improve' our biological make-up, so as to produce human beings much superior to us. There are, however, good reasons why any sort of intended amelioration of mankind's genetic make-up may not be wise. We can advance human evolution much more rapidly and effectively through cultural adaptation.

Beyond biology: culture

Humans are notably different from other animals not only in anatomy, but also, and no less importantly, in their behaviour, both as individuals and socially. Distinctive human behavioural traits include our exalted intelligence: language; self-awareness and death awareness; tool-making and technology; science, literature and art; ethics and religion; social organization and cooperation; legal codes and political institutions (see *Am I Really a Monkey?*).

A distinctive human social trait is culture, understood in a broad sense as the set of non-biological human activities and creations. Culture is transmitted by instruction and learning, by example and imitation, and by any other means of communication. Culture is acquired by every person from parents, relatives and neighbours, and from all of humankind.

As pointed out in *Am I Really a Monkey?* cultural inheritance makes possible the cumulative transmission of experience through the generations. Animals do not transmit their experience from generation to generation. Humans have developed cultures because they can transmit cumulatively their experiences from generation to generation.

Organisms adapt to the environment by means of natural selection, by changing over generations their genetic constitution to suit the demands of the environment. But humans, and humans alone, can also adapt by changing the environment to suit the needs of their genes. Cultural adaptation is more effective than biological adaptation because it is more rapid, because it can be directed and because it is cumulative from generation to generation. Whenever a need arises, humans can directly pursue the appropriate cultural changes to meet the challenge. Personal computers, cellphones

> *Cultural adaptation is more effective than biological adaptation because it is more rapid, because it can be directed and because it is cumulative from generation to generation.*

and the internet were invented and developed to meet perceived needs in human communication and information processing. The cultural innovations that humans achieve in any generation are accumulated to the cultural adaptations of the past and will accumulate with the cultural innovations of the future. Personal computers emerged after the invention and development of mainframe computers. The broad benefits of the internet and search engines such as Google and Yahoo! depend on the existence of personal computers. The benefits of cellphones became obvious only after the extensive use of traditional telephones for several decades.

Biology to culture

Cultural adaptation, if wisely used by humankind, will surely improve health and increase length of life, but it should also yield new lifestyles, individual and sociopolitical, that could be extremely rewarding. With the advanced development of the human brain, biological evolution has transcended itself, opening up a new mode of evolution: adaptation by technological manipulation of the environment. Only humans (at least to any significant degree) have developed the capacity to adapt to hostile environments by modifying the environment according to the needs of their genes. It is the human brain (or rather, the human mind) that has made us the most successful, by most meaningful standards, living species. It is cultural adaptation, much more than genetic engineering, that holds the promise of a better world for humankind.

CAN I CLONE MYSELF?
Genes can be cloned; a person cannot

*T*he genetic make-up of a person can be cloned; the individual
cannot. An individual is the outcome of the interactions between
that individual's genotype and the environment. The genetic make-
up of a person is fixed at conception, made up of the two sets of
genes received, one from the father and the other from the mother.
The environment includes all the experiences to which that genotype
has been exposed from conception to death, from the mother's womb
through childhood, schooling, family and society.

The eminent geneticist and Nobel Laureate George W. Beadle
wrote several decades ago:

> *Few of us would have advocated preferential multiplication of
> Hitler's genes. Yet who can say that in a different cultural context
> Hitler might not have been one of the truly great leaders of men, or
> that Einstein might not have been a political villain.*

Seeking to clone the Einsteins, the Lincolns and the
Gandhis, we might obtain instead Stalins, Hitlers and Rasputins.
Cloning the genes I received at conception from my father and
mother would result in a person who might resemble me in
appearance, but would surely be quite different with respect to
what counts the most, what is sometimes encompassed by words
such as 'personality', 'character' and the like.

Cloning: genes, cells, individuals

Biologists use the term 'cloning' with variable meanings, although all uses imply obtaining more or less precise copies of a biological entity. Three common uses refer to cloning genes, cloning cells and cloning individuals. Cloning an individual, particularly in the case of a multicellular organism such as a plant or an animal, is not strictly possible. The genes of an individual, the 'genome', can be cloned, but the individual cannot. This is the key statement when responding to the question 'Can I clone myself?'

Cloning genes – or, more generally, cloning DNA segments – is routinely done in many genetics laboratories throughout the world. A favoured and extensively used technology is PCR (polymerase chain reaction), invented in the 1990s by Kary Mullis, who received the Nobel Prize in recognition. With the PCR technique it is possible to obtain billions of virtually identical copies of a gene or DNA segment in just a few hours. This enormous multiplication of a DNA segment provides geneticists with enough material to investigate its nucleotide sequence and other properties.

> *Cloning an individual, particularly in the case of a multicellular organism such as a plant or an animal, is not strictly possible.*

Technologies for cloning cells in the laboratory are even older, some nearly seven decades old; they are used for reproducing a particular type of cell, for example a skin or a liver cell, in order to investigate its characteristics. Cell cloning is a natural process in two obvious ways. First, in the case of bacteria and other microorganisms that reproduce by cloning; that is, the splitting of an individual cell into two cells that are more or less identical to one another and to the mother cell. Cell cloning, or cell duplication, also occurs in multicellular organisms such as plants or animals, as cells multiply while making up a particular tissue, skin or muscle or red blood cells. Multicellular organisms start as one cell that duplicates again and again, although the process includes differentiation of the cell types that make up different tissues. In development, some cells duplicate exactly, or nearly exactly, as is the case, for example, with skin cells or red blood cells, while others differentiate during their replication. Thus, human embryonic stem cells develop into epithelial cells, muscle cells and the more than 200 other types of cells that exist in humans.

Individual cloning occurs naturally in the case of identical twins, when two individuals develop from a single fertilized egg. These twins are called 'identical' precisely because they are genetically identical to each other. The birth of identical twins is a relatively rare event in humans, but is frequent in some animals such as armadillos – whose litters consist of genetically identical quadruplets – some insects and others.

The sheep Dolly was acknowledged in 1997 by the press and the public as the first successful instance of the artificial cloning of a multicellular animal. In fact Dolly was the first *mammal* artificially cloned using an adult cell as the source of the genotype. Frogs and other amphibians had been obtained by artificial cloning as early as 50 years before.

Cloning an animal proceeds as follows. First, the genetic information in the ovules of a female is removed or neutralized. Somatic (i.e. body) cells are taken from the individual selected to be cloned, and the cell nucleus (where the genetic information is stored) is transplanted to the neutralized ovules. Then the ovules so 'fertilized' are stimulated to start their embryonic development. It is thus possible to obtain numerous individuals that are genetically identical to the donor; that is, individuals that from a genetic perspective are identical twins of the donor and of one another.

A human individual consists of about one trillion cells, and a piece of skin may have millions of cells. Theoretically, one could extract the genetic material of hundreds or thousands of cells from a small piece of skin and implant it in each of hundreds or thousands of genetically neutralized ovules, thus obtaining a multitude of individuals genetically as similar to one another and to the donor as two identical twins are. No sane person would propose to proceed in the described manner with a human, but the procedure could be carried out with, say, a cow successful in producing large quantities of milk, thereby obtaining an economically valuable herd.

Human cloning has occasionally been suggested as a way to improve the genetic endowment of mankind, by cloning individuals of great achievement – for example, in sport, music, the arts, science, literature or politics – or acknowledged virtue. These suggestions seemingly have never been taken seriously. But some individuals have expressed a wish, however unrealistic, to be cloned, and some physicians have on occasion advertised that they were ready to carry out the cloning. The obstacles and drawbacks are many and insuperable, at least at the present state of knowledge and relevant technology.

Can a human be cloned?

The correct answer is 'no'. What is cloned in any case are the genes, not the individual; the genotype, not the phenotype. But the technical obstacles are immense even for cloning a human's genotype.

Ian Wilmut, the British scientist who directed the Dolly cloning project, succeeded only after 270 trials. The rate of success for cloning mammals has notably increased over the years without ever reaching 100 per cent. After several years of effort, Wilmut wrote that 'Our survival rates are still very low [less than 4 per cent] and most pregnancy failures occur just before term, which would be devastating and cruel for humans.' The animals cloned to the present include mice, rats, goats, sheep, cows, pigs, horses and other mammals. In all instances it seems that the great majority of pregnancies end in spontaneous abortion. Moreover, as Wilmut noted, in many cases the death of the foetus occurs close to term, which would have devastating health and emotional consequences in the case of humans.

In mammals in general, the limited number of successful births do not grow into healthy individuals. It seems that to date in the immense majority, perhaps in all cases, the animals produced by cloning suffer from serious health handicaps; among others, gross obesity, early death, distorted limbs and dysfunctional immune systems and organs, including liver and kidneys. Even Dolly had to be euthanized after only a few years of life, because her health was rapidly decaying. As Wilmut stated in 2001: 'Those that survive often display respiratory distress and circulatory problems. Even apparently healthy survivors may suffer immune dysfunction, or kidney or brain malformation.'

The causes of such deficiencies are not well known. In 2002 it was shown that out of 10,000 genes analyzed in the placentas and livers of mice obtained by cloning, 400 genes were functioning badly. The low rate of cloning success may improve in the future. Human cloning would still face ethical objections from a majority of concerned people, as well as opposition from diverse religions. Moreover, there remains the theme repeated throughout this essay: it might be possible to clone a person's genes, but the individual cannot be cloned. The character, personality and features other than anatomical and physiological that make up the individual are not determined by the genotype.

Personhood: genotype versus phenotype

The genetic make-up of an individual is called 'genotype'. The 'phenotype' refers to what the individual is, which includes not only external appearance and anatomy, but also physiology, as well as behavioural predispositions and attributes, which in the case of humans are all-important, encompassing intellectual abilities, moral and religious values, aesthetic preferences and, in general, all other behavioural characteristics or features, acquired by imitation, learning or in any other way throughout the individual's life. The genotype contributes to the phenotype, but in humans more than in any other organisms, it does not strictly determine it. The phenotype results from complex networks of interactions between different genes, and between the genes and the environment.

All life experiences of a human being, conscious or not, influence what the person turns out to be. A person's environmental

All life experiences of a human being, conscious or not, influence what the person turns out to be.

influences begin, importantly, in the mother's womb and continue after birth, through childhood, adolescence and the whole life. Impacting behavioural experiences are associated with family, friends, schooling, social and political life, reading, aesthetic and religious experiences and every other event in the person's life.

The genotype of a person has an unlimited number, virtually infinite, of possibilities, only some of which – no matter how enormously diverse – will be experienced in a particular individual's lifetime. Necessarily, the life experiences will always vary from one to another person, even in the case of identical twins. Numerous investigations have shown that identical twins differ in many behavioural and even physiological traits, and that the differences increase with age, because the diversity of experiences increases. The behavioural differences are larger, not surprisingly, among twins who are adopted and grow up in different families. The diversity of experiences would be very large and unpredictable in the case of 'identical twins' raised in different generations. A person's genome, if it were cloned, might be thought of as a genetically identical twin, but the disparate life circumstances experienced one generation later would surely result in a very different individual, even if anatomically they would resemble the genome's donor at a similar age.

An illustration of environmental effects on the phenotype, and of interactions between the genotype and the environment, is shown in the figure opposite. Three plants of the cinquefoil *Potentilla glandulosa* were collected in California – one on the coast at about 100 feet above sea level, the second at about 4,600 feet and the third in the Alpine zone of the Sierra Nevada at about 10,000 feet. Three cuttings of each plant were planted in three experimental gardens at different altitudes, using the same gardens for all three plants. The division of one plant ensured that all three parts planted at different altitudes had the same genotype; that is, they were genetic clones from one another. (*P. glandulosa*, like many other plants, can be reproduced by 'cuttings', which are genetically identical.)

Comparison of the plants in any row shows how a given genotype gives rise to different phenotypes in different environments. Genetically identical plants (for example, those in the bottom row) may prosper or die depending on the environmental conditions. Besides the obvious divergence in appearance, there are differences in fertility, growth rates, etc.

Plants from different altitudes are known to be genetically different. Hence, comparison of the plants in any column shows that in a given environment different

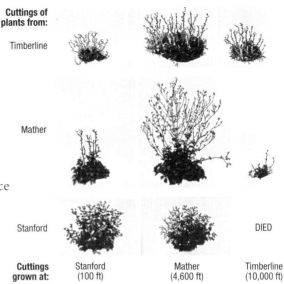

Cuttings of plants from:			
Timberline			
Mather			
Stanford			DIED
Cuttings grown at:	Stanford (100 ft)	Mather (4,600 ft)	Timberline (10,000 ft)

INTERACTING EFFECTS OF THE GENOTYPE AND THE ENVIRONMENT ON THE PHENOTYPE (APPEARANCE). CUTTINGS FROM THREE PLANTS OF THE CINQUEFOIL *POTENTILLA GLANDULOSA*, COLLECTED AT DIFFERENT ALTITUDES IN CALIFORNIA, WERE PLANTED EACH IN ONE OF THREE EXPERIMENTAL GARDENS.

genotypes result in different phenotypes. An important observation derived from this experiment is that there is no single genotype that is 'best' in all environments. For example, the plant from near sea level that prospers there fails to develop at 10,000 feet. Likewise, the plant collected at 10,000 feet prospers at that altitude but withers at sea level.

The interaction between the genotype and the environment is similarly significant in the case of animals. In one experiment conducted several years ago, two strains of rats were selected; one for brightness at finding their way through a maze and the other for dullness. Selection was done in the bright strain by using the brightest rats of each generation to breed the following generation, and in the dull strain by breeding the dullest rats every generation. After many generations of selection, the descendant bright rats made only about 110 errors running through the maze, whereas dull rats averaged 165 errors. That is a 50 per cent difference. If we'd been thinking in terms of human IQ, the difference in performance between the two strains of rats would be comparable to a difference between IQ scores of 80 and more than 120. However, the differences disappeared when rats of both strains were raised in an unfavourable environment of severe deprivation, where both strains averaged 170 errors. A comparable change in IQ score for individuals raised in a deprived environment would be a reduction from 120 to 80. The differences also nearly disappeared when the rats were raised with abundant food and other unusually favourable conditions. As with the cinquefoil plants, we see both that a given genotype gives rise to different phenotypes in different environments, and that the differences in phenotype between two genotypes change from one environment to another.

Cloning humans?

In the second half of the twentieth century, as dramatic advances were taking place in genetic knowledge, as well as in the technology often referred to as 'genetic engineering', some utopian proposals were advanced, at least as ideas that should be explored and considered as possibilities once the technologies had sufficiently progressed. These proposals suggested that persons of great intellectual or artistic achievement or great virtue be cloned. If this were accomplished in large numbers, the genetic constitution of mankind would, it was argued, improve considerably. Such utopian proposals are grossly misguided. Seeking to multiply great benefactors of humankind, persons of great intelligence or character, we might obtain instead the likes of Stalin, Hitler or bin Laden. There is no reason whatsoever to expect that cloning the genomes of individuals with excellent attributes would produce individuals similarly endowed with virtue or intelligence. Identical genomes yield, in different environments, individuals who may be quite different. Environments cannot be reproduced, particularly several decades apart.

Are there circumstances that would justify cloning a person, because he or she wants it? One might think of a couple unable to have children, or a man or woman who does not want to marry, or of other special cases. It must first be pointed out that technology has not yet been developed to an extent that would make it possible to produce a healthy human individual by cloning. Second, and most important, the individual produced by cloning would be a very different person from the one whose genotype is cloned, as emphasized above.

There is no reason whatsoever to expect that cloning the genomes of individuals with excellent attributes would produce individuals similarly endowed with virtue or intelligence.

Ethical, social and religious values will come into play when seeking to decide whether a person might be allowed to be cloned. Most people are likely to disapprove. Indeed, many countries have prohibited human cloning. In 2004, the issue of cloning was raised in several countries where legislatures were also considering whether research on embryonic stem cells should be supported or allowed. On 12 March 2004, the Canadian parliament passed legislation permitting research with stem cells from embryos under specific conditions, but human cloning was banned, and the sale of sperm and payments to egg donors and surrogate mothers was prohibited. On 9 July 2004, the French parliament adopted a new bioethics law that allows embryonic stem cell research but considers human cloning a 'crime against the human species'. Reproductive cloning experiments would be punishable by up to 20 years in prison. Japan's Cabinet Council for Science and Technology Policy voted on 23 July 2004 to adopt policy recommendations that would permit the limited cloning of human embryos for scientific research, but not the cloning of individuals.

One anticipated outcome of cloning research with embryonic cells is, in Japan as elsewhere, the cloning of organs. The optimal 'donor' for a person in need of a kidney transplant would be a kidney cloned from the genome of the patient. If cells from a person are cloned so that they differentiate into a kidney, a liver or some other organ intended to replace a sick organ from the donor, it seems likely that governments might not object and moralists might approve.

WHERE DOES MORALITY COME FROM?
Biology versus culture

'*A*ny animal whatever, endowed with well-marked social instincts, would inevitably acquire a moral sense or conscience, as soon as its intellectual powers had become as well developed, or nearly as well developed, as in man ... I do not wish to maintain that any strictly social animal, if its intellectual faculties were to become as active and as highly developed as in man, would acquire exactly the same moral sense as ours .... [T]hey might have a sense of right and wrong, though led by it to follow widely different lines of conduct.'[5]

Darwin is affirming here that the moral sense, or conscience, is a necessary consequence of high intellectual powers, such as exist in modern humans. Therefore, since human intelligence is an outcome of natural selection, the moral sense would be an outcome of natural selection too. He further implies that the moral sense is not by itself directly promoted by natural selection, but only indirectly as a necessary consequence of high intellectual powers, which are the attributes that natural selection *is* directly promoting. Moreover, according to Darwin, having a moral sense does not by itself determine what the moral norms would be: which sorts of actions might be sanctioned and which ones condemned.

Moral behaviour versus moral norms
Darwin's distinction between the moral sense or conscience on the one hand and the norms that guide the moral sense or conscience

on the other is fundamental. It is a distinction I will elaborate. Much of the post-Darwin historical controversy, particularly between scientists and philosophers, as to whether the moral sense is or is not biologically determined has arisen owing to a failure to make that distinction. Scientists often affirm that morality is a human biological attribute because they are thinking about the 'moral sense', the predisposition to make moral judgements: that is, to judge some actions as good and others as evil. Some philosophers argue that morality is not biologically determined, but rather comes from cultural traditions or from religious beliefs, because they are thinking about moral codes, the sets of norms that determine which actions are judged to be good and which evil. They point out that moral codes vary from culture to culture and, therefore, are not biologically predetermined.

Aristotle and other philosophers of classical Greece and Rome, as well as many other philosophers throughout the centuries, held that humans hold moral values by nature. A human is not only *Homo sapiens*, but also *Homo moralis*. But biological evolution brings about two important issues: timing and causation. We do not attribute ethical behaviour to animals (not to all animals, and not

> *Darwin's distinction between the moral sense or conscience on the one hand and the norms that guide the moral sense or conscience on the other is fundamental.*

to the same extent as to humans in any case). When did ethical behaviour come about in human evolution? Did modern humans have an ethical sense from the beginning? Did Neanderthals hold moral values? What about the ancestral species *H. erectus* and *H. habilis*? And how did the moral sense evolve? Was it directly promoted by natural selection? Or did it come about as a by-product of some other attribute (such as intelligence) that was the direct target of selection? Alternatively, is the moral sense an outcome of cultural evolution rather than of biological evolution?

The question whether ethical behaviour is biologically determined may, indeed, refer to either one of the following two issues. First, is the capacity for ethics – the moral sense, the

169

proclivity to judge human actions as either right or wrong – determined by the biological nature of human beings? Second, are the systems or codes of ethical norms biologically determined? A similar distinction can be made with respect to language. The question whether the capacity for symbolic creative language is determined by our biological nature is different from the question whether the particular language we speak – English, Spanish, Chinese, etc. – is biologically determined, which in the case of language obviously it is not.

The first question asks whether or not the biological nature of humans is such that we are necessarily inclined to make moral judgements and to accept ethical values, to identify certain actions as either right or wrong. It asks whether moral behaviour is an outcome of biological evolution. Affirmative answers to this first question do not necessarily determine what the answer to the second question should be. Independent of whether or not humans have a biologically determined moral sense, it remains to be determined whether particular moral prescriptions are in fact determined by the biological nature of humans, or whether they are chosen by society, or by individuals. The need for moral values does not necessarily tell us what the moral values should be, in the same way that the capacity for language does not determine which language we will speak.

> *The need for moral values does not necessarily tell us what the moral values should be.*

I will propose that humans are ethical beings by their biological nature; that humans evaluate their behaviour as either right or wrong, moral or immoral, as a consequence of their eminent intellectual capacities, which include self-awareness and abstract thinking. These intellectual capacities are products of the evolutionary process, but they are distinctively human. I will add that the moral norms according to which we evaluate particular actions as either morally good or morally bad (as well as the grounds that may be used to justify the moral norms) are products of cultural, not biological evolution.

Evolution of the moral sense

Why is the moral sense an attribute determined by our biology, and hence an attribute that came about by natural selection? The moral sense refers to the evaluation of some actions as virtuous, or morally good, and others as evil, or morally bad. Morality in this sense is the urge or predisposition to judge human actions as either right or wrong in terms of their consequences for other human beings.

In this sense, humans are moral beings by nature because their biological constitution determines the presence in them of the three necessary conditions for ethical behaviour. These conditions are: the ability to anticipate the consequences of one's own actions; the ability to make value judgements; and the ability to choose between alternative courses of action. These abilities exist as a consequence of the eminent intellectual capacity of human beings.

The ability to anticipate the consequences of one's own actions is the most fundamental of the three conditions required for ethical behaviour. Only if I can anticipate that pulling the trigger will shoot the bullet, which in turn will strike and kill my enemy, can the action of pulling the trigger be evaluated as nefarious. Pulling a trigger is not in itself a moral action; it becomes so by virtue of its relevant consequences.

My action has an ethical dimension only if I anticipate these consequences. The ability to anticipate the consequences of one's actions is closely related to the ability to establish the connection between means and ends; that is, of seeing a means precisely as a means, as something that serves a particular end or purpose. This ability to establish the connection between means and their ends requires the ability to anticipate the future and to form mental images of realities not present or not yet in existence.

The ability to establish the connection between means and ends happens to be the fundamental intellectual capacity that has made possible the development of human culture and technology. The remote evolutionary roots of this capacity may be found in the evolution of bipedalism, which transformed the anterior limbs of our ancestors from organs of locomotion into organs of manipulation. The hands thereby gradually became organs adept for the construction and use of objects for hunting and other activities that improved survival and reproduction; that is, which increased the reproductive fitness of their carriers. The construction of tools depends not only on manual dexterity, but on perceiving them precisely as tools, as objects that help to perform certain actions; that is, as means that serve certain ends or purposes: a knife for cutting, an arrow for hunting, an animal skin for protecting the body from the cold. Natural selection promoted the intellectual capacity of our bipedal ancestors because increased intelligence facilitated the perception of tools as tools, and therefore their construction and use, with the ensuing amelioration of biological survival and reproduction.

The development of the intellectual abilities of our ancestors took place over many millennia, gradually increasing the ability to connect means with their ends and, hence, the capacity of making ever more complex tools serving more remote purposes. The ability to anticipate the future, essential for ethical behaviour, is therefore closely associated with the development of the ability to construct tools, an ability that has produced the advanced technologies of modern societies and is largely responsible for the success of humans as a biological species.

The second condition for the existence of ethical behaviour is the ability to make value judgements, to perceive certain objects or deeds as more desirable than others. Only if I can see the death of my enemy as preferable to his survival (or vice versa) can the action leading to his demise be thought of as moral. If the consequences of alternative actions are neutral with respect to value, an action cannot be characterized as ethical. Values are of many sorts: not only ethical, but also aesthetic, economic, gastronomic, political and so on. But in all cases, the ability to make value judgements depends on the capacity for abstraction; that is, on the capacity to perceive actions or objects as members of general classes. This makes it possible to compare objects or actions with one another and to perceive some as more desirable than others. The capacity for abstraction requires an advanced intelligence such as exists in humans and apparently in them alone.

The third condition necessary for ethical behaviour is the ability to choose between alternative courses of action. Pulling the trigger can be a moral action only if one has the option not to pull it. A necessary action beyond conscious control is not a moral action: the circulation of the blood or the digestion of food are not moral actions. Whether there is free will is a question much discussed by philosophers, and the arguments are long and involved. I advance two considerations that are common-sense evidence of the existence of free will. One is personal experience, which indicates that the possibility to choose between alternatives is genuine rather than only apparent. The second consideration is that when we confront a given situation

> *Only if I can see the death of my enemy as preferable to his survival (or vice versa) can the action leading to his demise be thought of as moral.*

that requires action on our part, we are able mentally to explore alternative courses of action, thereby extending the field within which we can exercise our free will. In any case, if there were no free will, there would be no ethical behaviour; morality would only be an illusion. The point is that free will is dependent on the existence of a well-developed intelligence, which makes it possible to explore alternative courses of action and to choose one or another in view of the anticipated consequences.

Is morality adaptive?

Two issues concerning the explanation of moral behaviour just developed are: (1) is morality an adaptation directly favoured by natural selection rather than simply a by-product of high intelligence?; and (2) does morality occur in other animals, either as directly promoted by natural selection or as a consequence of animal intelligence, even if only as a rudiment?

> *The norms of morality must be consistent with biological nature, because ethics can only exist in human individuals and in human societies.*

The answer to the first question is negative. Morality consists of judging certain actions as either right or wrong; not of choosing and carrying out some actions rather than others, or evaluating them with respect to their practical consequences. It seems unlikely that making moral judgements would promote the reproductive fitness of those judging an action as good or evil. Nor does it seem likely that there might be some form of 'incipient' ethical behaviour that would then be further promoted by natural selection. The three necessary conditions for there being ethical behaviour are manifestations of advanced intellectual abilities.

It seems rather that the target of natural selection was the development of these advanced intellectual capacities. This was favoured by natural selection because the construction and use of tools improved the strategic position of our biped ancestors. Once bipedalism evolved, and after tool-using and tool-making became practical, those individuals more effective in these functions had a greater probability of biological success. The biological advantage provided by the design and use of tools persisted long enough so that intellectual abilities continued to increase, eventually yielding the eminent development of intelligence that is characteristic of *Homo sapiens*.

Whence moral codes?

I have distinguished between moral behaviour – judging some actions as good, others as evil – and moral codes – the precepts or norms according to which actions are judged. Moral behaviour, I have proposed, is a biological attribute of *H. sapiens*, because it is a necessary consequence of our biological make-up, namely our high intelligence. Moral codes, however, are products not of biological evolution, but of cultural evolution.

It must first be stated that moral codes, like any other cultural systems, cannot survive for long if they run in outright contrast to our biology. The norms of morality must be consistent with biological nature, because ethics can only exist in human individuals and in human societies. One might therefore also expect – and indeed it is the case – that accepted norms of morality will promote behaviours that increase the biological fitness of those who behave according to them, such as child care. But this is neither necessary nor indeed always the case: some moral precepts common in human societies have little or nothing to do with biological fitness, and some are contrary to it.

Before going any further, it seems worthwhile to consider briefly the proposition that the justification of the codes of morality derives from religious convictions and only from them. There is no necessary or logical connection between religious faith and moral principles, although there is usually a motivational or psychological connection. Religious beliefs explain why people accept particular ethical norms, because they are motivated to do so by their religious convictions. But in following the moral dictates of one's religion, one is not rationally justifying the moral norms that one accepts. It may, of course, be possible to develop such rational justification; for example, when a set of religious beliefs contains propositions about human nature and the world from which a variety of ethical norms can be logically derived. Indeed, religious authors, including Christian theologians, do often propose to justify their ethics on rational foundations concerning human nature. But in this case, the logical justification of the ethical norms does not come from religious faith as such, but from a particular conception of the world; it is the result of philosophical analysis grounded on certain premises.

It may well be that the motivational connection between religious beliefs and ethical norms is the decisive one for the religious believer. But this is true in general: most people, religious or not, accept a particular moral code for social reasons, without trying to justify it rationally by means of a theory from which the moral norms can be logically derived. They accept the moral codes that prevail in their societies, because they have learned such norms from parents, school, religion or other authorities.

> *Cultural heredity does not depend on biological inheritance, from parents to children, but is transmitted also horizontally and without biological bounds.*

The question therefore remains: how do moral codes come about?

The short answer is, as already stated, that they are products of cultural evolution, a distinctive human mode of evolution that has surpassed the biological mode, because it is faster and because it can be directed. Cultural evolution is based on cultural heredity, which is Lamarckian, rather than Mendelian, so that acquired characteristics are transmitted. Most importantly, cultural heredity does not depend on biological inheritance, from parents to children, but is transmitted also horizontally and without biological bounds. A cultural mutation, an invention (think of the laptop computer, the cellphone or rock music), can be extended to millions and millions of individuals in less than one generation.

Since time immemorial, human societies have experimented with moral systems. Some have succeeded and spread widely through humankind, like the Ten Commandments, although other moral systems persist in different human societies. Many moral systems of the past have become extinct because they were replaced or because the societies that held them died out. The moral systems that currently exist in humankind are those that were favoured by cultural evolution. They were propagated within particular societies for reasons that might be difficult to fathom, but which surely must have included the perception by individuals that they were beneficial, at least to the extent that they promoted social stability and success. Empathy, or the predisposition to mentally assimilate the feelings of other individuals, is a behavioural predisposition favouring altruism and other common moral patterns of behaviour. Acceptance of some precepts is reinforced in many societies by civil authority (e.g. those who kill or commit adultery will be punished) and by religious beliefs (God is watching and you'll go to hell if you misbehave). Legal and political systems as well as belief systems are themselves outcomes of cultural evolution.

Darwin's moral optimism

In Chapter V of *The Descent of Man*, entitled 'On the Development of the Intellectual and Moral Faculties during Primeval and Civilized Times', Darwin writes:

> *There can be no doubt that a tribe including many members who, from possessing in a high degree the spirit of patriotism, fidelity, obedience, courage, and sympathy, were always ready to give aid to each other and to sacrifice themselves for the common good, would be victorious over most other tribes; and this would be natural selection. At all times throughout the world tribes have supplanted other tribes; and as morality is one element in their success, the standard of morality and the number of well-endowed men will thus everywhere tend to rise and increase.*

Darwin is making two important assertions. First, that morality may contribute to the success of some tribes over others, which is natural selection in the form of group selection (that is, selection that favours a group or population relative to others, without necessarily favouring some individuals over others in the same group or population). Second, that standards of morality will tend to improve over human history precisely on grounds of group selection, because the higher the moral standards of a tribe, the more likely its success. This assertion of moral optimism depends on which standards are thought to be 'higher' than others. If the higher standards are defined by their contribution to the success of the tribe, then the assertion is circular. But in Darwin's view, there are particular standards that would contribute to tribal success: patriotism, fidelity, obedience, courage and sympathy.

Parental care is a behaviour generally favoured by natural selection that may be present in virtually all codes of morality, from primitive to more advanced societies. There are other human behaviours sanctioned by moral norms that have biological correlates favoured by natural selection. One example is monogamy, which occurs in some animal species but not many. It is also sanctioned in many human cultures, but surely not in all. Polygamy is accepted in some current human cultures and was more so in the past. Food sharing outside the mother–offspring unit rarely occurs in primates, with the exception of chimpanzees – and, apparently, in capuchin monkeys – although even in chimpanzees it is highly selective and often associated with reciprocity. A more common form of mutual aid among primates is coalition formation; alliances are formed in fighting other conspecifics, although these alliances are labile, with partners readily changing sides.

The norms of morality, as they exist in any particular human society or culture, are felt to be universal within that culture. Yet, like other elements of culture, they are continuously evolving, often within a single generation. For example, Western societies have recently experienced the moralization and amoralization of diverse behaviours: smoking has become moralized and is now treated as immoral. Other behaviours, such as divorce, illegitimacy and homosexuality, have become amoralized, switched from moral failings to lifestyle choices. The legal and political systems that govern human societies, as well as the belief systems held by religion, are themselves outcomes of cultural evolution, as it has eventuated throughout human history, particularly over the last few millennia.

IS LANGUAGE A UNIQUELY HUMAN ATTRIBUTE?

Humans speak; ants and bees communicate

'*Articulate language is, however, peculiar to man; but he uses in common with the lower animals inarticulate cries to express his meaning, aided by gestures and movements of the muscles of the face. This especially holds good with the more simple and vivid feelings, which are but little connected with our higher intelligence ... It is not the mere power of articulation that distinguishes man from other animals ... but it is his large power of connecting definite sounds with definite ideas; and this obviously depends on the development of the mental faculties.*'[6]

Language is one of our species' most distinctive behavioural traits. No other animal speaks like we do, by means of symbolic creative languages. When dealing with the evolution of language, we are faced with the question of how and when this evolution took place. The faculty of language requires a prior substrate available only to humans: advanced intelligence as it exists in *Homo sapiens*, and only in *Homo sapiens* among living species.

Communication versus language

The following is a dictionary definition of language: 'the words, their pronunciation, and the methods of combining them used and understood by a community'. Language so defined exists only in humans. But a main purpose of language is communication between individuals. The same dictionary defines communication:

'a process by which information is exchanged between individuals through a common system of symbols, signs, or behaviour'. Communication occurs between animals, though not by means of language. An example given by the same dictionary is: 'the function of pheromones in insect communication'.

Animals communicate by means of sounds, gestures, chemical substances and body movements. But language is a distinctive possession of human beings. No other species has conventionalized its utterances so that they constitute a systematic symbolism in the way that language does. Human language is creative, virtually unlimited in scope, in two ways: the number of words that can be created, and the ways in which words can be combined in phrases, sentences, paragraphs and so on.

A sophisticated system of animal communication is the 'language of the bees'. When a foraging worker bee discovers a food source or a desirable new nest site, she returns to the hive and indicates the location to other workers by performing a waggle dance, a flight patterned as a figure-of-eight repeated over and over again among the surrounding crowd of sister workers. She rapidly waggles the tip of her abdomen and emits a buzzing sound by vibrating her wings. She is conveying information about the direction of the food source or nest site relative to the position of the sun, and about the distance: the farther away the target, the longer the duration of the dance. The waggle dance, like all other forms of non-human communication, is severely limited in comparison with human verbal language. The rules are genetically fixed and designate direction and distance.

> *The faculty of language requires a prior substrate available only to humans: advanced intelligence as it exists in* Homo sapiens, *and only in* Homo sapiens *among living species.*

Other known modes of communication in insects and other lower animals are always stereotyped. Each of a limited number of signals is associated with a fixed response. Chemical signals are the most widespread mode of communication among insects. The female silkworm attracts males by emitting minute quantities of a complex alcohol at the tip of her abdomen. The male catches the molecules in thousands of distinctive hairs on each of his two antennae, which respond to no other molecules than the alcohol emitted by female silkworm moths.

Ants, termites and other social insects mostly communicate by pheromones, species-specific chemical molecules. Tactile cues are also used, in addition to the chemical signals. Non-social insects also use chemical and tactile communication systems, which may be quite complex, as in crickets for example.

The most extensive means of communication in fishes, birds and mammals are displays, complex patterns of behaviour that have evolved to convey specific information. Display behaviours are most often used to achieve attraction between females and males in association with mating. Display patterns and the associated responses have been studied in fishes (for example, guppies, sticklebacks and sunfishes), birds (ducks, herons, gulls, sparrows and the great tit) and mammals (deer, elks, gazelles, lemurs, tamarins, rhesus monkeys and numerous other primates, including the great apes).

Animals also communicate by sounds, with more or less elaborate songs, alarm calls, threatening growls and grunts. Elaborate songs are found in numerous bird species, but also in whales and other marine mammals. Crickets communicate by persistent loud 'songs' that are species-specific. Most communication signals in non-human animals are genetically fixed. But the great apes and other primates may occasionally learn additional gestures or growls. Some species of birds are able to sing their specific songs even before they have heard them from their parents or other conspecifics. In other species, the young are able to sing their specific song only after hearing it from their parents or other members of the species. Experiments have shown that in some bird species the young will learn the first song they hear, whether or not it comes from conspecifics, although in nature, the first song they'll hear will probably be their parents'.

The origin of language

True syntax, as it occurs in human linguistics, where the meaning of combinations of signals depends on their order of appearance, and the signals combine and their order varies without bounds, is never present in animal communication. Non-human animals totally lack the two virtually infinite modes of articulation – words and word combinations – that characterize the creative features of human languages.

Any explanation for the evolution of human speech faces the so-called 'Plato's problem' or 'poverty of stimulus'. In a very short time, and based on dispersed and confused information, children manage to understand and produce correct syntactic constructions. The American linguist Noam Chomsky has postulated the existence of genetic baggage that makes the capacity to speak an innate asset of our species. This faculty would be a product of evolution, as a distinctively human trait.

According to Chomsky, the predisposition to language is innate, in the same way that the capacity to produce and interpret gestures or recognize faces is innate. This does not imply that any kind of sign communication system would have appeared in evolution due to the same mechanisms and at the same time as linguistic competence. Rather, the path to language is the combination of a very diverse series of communication aptitudes whose evolution may have extended over two million years. But it seems that at a certain point in the evolution of our ancestors, communicative capacities took a completely new direction owing to the appearance of three novelties: first, an organ that produced sounds capable of modulating vowels and consonants; second, means of phonetic/semantic identification which associates the combinations of vowels and consonants with meanings; and third, a combination of phonetic/semantic units capable of generating an unlimited number of messages subjected to syntactic rules. Such a combination of capacities is exclusively human. It may have been generated by very specific mutations that turned the previous communicative abilities into a new and unique kind of language.

Genes and language

Linguistic competence is achieved rapidly and without systematic learning tasks. Hence it is necessary to posit the existence of some innate genetic predisposition. Which genes control this human capacity for speech?

FOXP2 was the first gene ever identified with a function related to language. It was isolated while studying a family with an inherited severe language and speech impairment. The affected members of the family had difficulties in selecting and sequencing fine orofacial movements, and exhibited grammatical deficits as well as slight non-verbal cognitive impairments. It was soon determined that the disorder was due to a point mutation in a gene situated on chromosome 7.

FOXP2 is not exclusively human. It is part of the genome of animals as evolutionarily distant as humans and mice, but only in humans is it associated with spoken language. The *FOXP2* protein is highly conserved among mammals: it has undergone only one amino-acid replacement during the 70 million years that elapsed between the last common ancestor of primates and mice and the last common ancestor of humans and chimpanzees. Since the divergence of the human and chimpanzee lineages 6–7 million years ago, the human protein has undergone two additional amino-acid changes, while the chimpanzee form has not changed. If the gene participates in laying down the neural circuits involved in speech and language, it seems possible that the last two mutations that occurred in the human lineage were crucial for the development of language.

> *Linguistic competence is achieved rapidly and without systematic learning tasks. Hence it is necessary to posit the existence of some innate genetic predisposition.*

Some authors have estimated that these events happened during the last 200,000 years. We do not know whether language would have been possible with a chimpanzee's or earlier hominid version of the gene. One possibility is that language appeared fairly rapidly in the hominid lineage with the advent of modern *H. sapiens*. However, the high degree of conservation of the protein, and of the pattern of the gene's expression in the brain, suggests that language and speech are, at least in part, supported by neural structures present in other species, which would indicate a gradual emergence of the capacity for language through the recruitment or fine-tuning of pre-existing neural pathways.

The vocal tract

Different human languages use different vocal expressions, although they all consist of an ordered succession of vowels and consonants. A significant issue is determining when hominins acquired the capability of pronouncing vowels and consonants in a way similar to modern humans. Chimpanzees are incapable of doing this, and thus it is reasonable to assume that this capacity was acquired during hominid evolution. But at what moment? Language does not fossilize, and written language does not appear until a few thousand years ago, the last split second of our species' history.

Language requires certain anatomical features, relative to the brain and the supralaryngeal vocal tract, the part of the throat that goes from the larynx to the oral cavity. The anatomical arrangement of the human supralaryngeal vocal tract allows a very particular kind of modulation of air flowing out. Through the coordination of the tongue, palate, teeth and lips we are able to pronounce a multitude of vowels and consonants. Vocalizing requires a larynx placed in a relatively low position, but in addition we need brain mechanisms suitable for sequencing the phonemes that make up words according to precise rules. Hence, it is possible that something helpful might be said about the evolution of language by studying the evolution of the necessary anatomical elements: the supralaryngeal vocal tract and the brain.

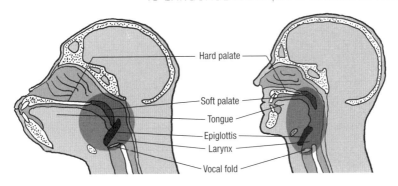

STRUCTURES OF THE VOCAL TRACTS (SHADED AREAS) IN CHIMPANZEES (LEFT) AND MODERN HUMANS (RIGHT). THE HUMAN VOCAL TRACT HAS EVOLVED IN WAYS THAT ALLOW US TO ENUNCIATE VARIOUS VOWELS AND CONSONANTS, WHICH IS NOT POSSIBLE FOR CHIMPANZEES.

Human speech sounds require that the length of the tube formed by the mouth be equivalent to that of another tube descending behind the tongue. Thus only a relatively low larynx will allow human vocalization. The reconstruction of the base of fossil hominin crania suggests that the larynx of australopiths was in a high position, similar to that of chimpanzees. The descent may have begun in *Homo erectus*. The study of the marks left by muscles on the basicrania, and the computational interpolation of chimpanzee and human anatomies, suggests that speech was a very late phenomenon, distinctive of anatomically modern humans. Some evidence indicates that the descent of the larynx did not take place until 40,000 years ago, as a second phase of the evolution of our species. The first phase, which began 200,000 years ago, would have modernized the cranium and, partially, the larynx, leading to a cavity about half of its current length and thus allowing an imperfect vocal behaviour. Only the second phase would lead to speech like ours.

The discovery of a hyoid bone – which is involved in the movements of the larynx – in Neanderthal fossils from the Mousterian site of Kebara, in the Near East, indicates, according to some authors, that the vocal apparatus of Neanderthals could have been similar to ours. However, most of the speculations on the possible language of Neanderthals are based on another kind of evidence: symbolic behaviour. We do not speak with our throat, but with our brain.

Evolution of the brain

Primate brain organization is fairly homogeneous. An Old World monkey, such as a macaque, is in this sense similar to a human being. But the human brain exhibits a conspicuous development of the temporal and pre-frontal areas related, precisely, to verbal communication and semantic processes. Determining how this evolution occurred would provide a solid base for speculations on the origin of language.

Examination of *Homo habilis* endocrania, and their comparison with those of *Australopithecus africanus*, shows that the Broca's and Wernicke's areas in the left hemisphere of the brain had already begun to develop in *A. africanus*. These areas are thought to be primarily responsible for the processes related to language. There is also an enlargement of the frontal lobes that seems to anticipate the increase in volume that occurred in later hominins. The palaeoanthropologist Phillip V. Tobias has suggested that a structural change of the brain that began with *A. africanus* would have consolidated with *H. habilis*.

What was the language of *H. habilis* like? Hardly anything can be said beyond the indications on endocranial casts. *H. habilis* may have possessed, in the best of cases, a language that was not quite modern. Tobias has proposed that we should not think of there being two phases in language evolution, one 'not completely human' and another 'completely human'. Rather he believes that the evolution of language was gradual, by means of a series of stages of increasing conceptual and syntactic complexity, together with a broadening of phonetic range. He argues that all these stages would qualify as 'spoken human language'.

Other authors, however, have noted two frontiers, so to speak, in the evolution of brain complexity. The first is the appearance of an 'essentially human' neurological organization in *H. habilis*. The second is the continuous and rapid increase in the encephalization quotient within the genus *Homo*.

The global frontal volume of the brain has not varied greatly throughout human evolution. But it is possible that some areas, such as the pre-frontal cortex, could have expanded in our species. This suggestion is compatible with the observation that current human, gorilla and chimpanzee endocranial measurements show the same relative size of the frontal lobe. However, there are significant differences favouring humans in certain sub-regions, suggesting that the human frontal lobe has undergone a process of reorganization in comparison with that of great African apes.

Brain costs

Brain tissue consumes a great amount of oxygen and glucose, and it does so continuously, independently of physical and mental states. This is true for all mammals, but the human brain's metabolic needs are enormous. It consumes 20 per cent of the oxygen required by the whole body, although it weighs only about 2 per cent as much as the body. Additionally, the metabolic index for the human cortex is 43 per cent greater than for the rest of the brain. These high metabolic requirements are an important consideration when assessing how natural selection may have favoured an expanding cortex. The only way to meet the brain's metabolic demands would have been a dietary change for the genus *Homo* towards richer nutrients, particularly meat. The intestine of modern humans is relatively smaller than the intestine of other primates. It would seem strange that throughout evolution, one organ – the brain – increasingly required more nutrients, while the intestines became reduced. The explanation is that the human digestive system is specialized, with a relatively long small intestine compared to the considerable length of the colon of apes. Herbivorous (vegetable-eating) primates, with larger intestines, have relatively smaller brains than frugivorous (fruit-eating) primates. Herbivorous diets require very large intestines for digestion, which would constitute a barrier for high encephalization.

A different explanation for the mechanism that allowed the evolutionary increase of brain size is, however, possible. The correlation between carnivorous diet and high encephalization is contradicted by the small brain in insectivorous bats, whose intestine is relatively smaller than that of bats that feed on fruits. The large brain/small intestine versus small brain/large intestine does not seem to be a general law. An alternative way to establish a relation between energetic resources and brain development is known as the maternal-energy hypothesis. This argues that the growth of a primate brain occurs initially within the foetus during pregnancy, and later during exterogestation with breastfeeding. Hence the energetic resources available to the mother while pregnant and breastfeeding are an essential variable. A mother can produce large-brained offspring by increasing the gestation period or the metabolic rate. Maternal energy could account for a 'passive' increase in encephalization. This would later be consummated by dietary changes – such as the increase in meat consumption – that allow for higher metabolic rates. The increase of the gestation period as it occurs in humans and the dietary change would jointly account for the large energy demands of the human brain.

Modern languages

Advanced symbolism indicates the presence of speech similar to ours, with writing as the last and most obvious example of symbolic communications. How was this capacity for symbolism acquired? It would seem that the cognitive capacities necessary to produce objects with a high symbolic content, such as art and other forms of representation, would not be very different from those required for speech.

There is archaeological and palaeontological evidence of early symbolic manifestations in Africa, dated more than 40,000 years ago. But the European (Aurignacian) artistic explosion 40,000–30,000 years ago (culminating later in the extraordinary artistic representations in caves such as Lascaux and Altamira) seems to have amounted to a gigantic qualitative and quantitative improvement.

If we rather see European Aurignacian culture as a revolution, there are several possibilities. One hypothesis suggests that there was some fairly rapid change in the brains of anatomically modern humans, which increased cognitive capacity. However, there is no incontrovertible direct evidence of such rapid transformation. We are still uncertain about the neural correlates of the so-called higher cognitive processes, such as aesthetic experience.

A behavioural hypothesis would argue that anatomically modern humans changed their behaviour when they became producers of an art that still amazes us today. The correlation of the artistic accomplishments with the behavioural habits of modern humans renders a hypothetical interpretation of the appearance of cave paintings and decorative objects. The transition from an occasional use of caves to their continuous occupation might be an explanation for why paintings on their ceilings appeared at a particular time. The permanent occupation of caves might be related to the climatic changes that occurred in Europe, culminating with the peak of the last and most extreme glaciation around 16,000 years ago, which coincides with the dates of the great cave paintings.

> *It would seem that the cognitive capacities necessary to produce objects with a high symbolic content, such as art and other forms of representation, would not be very different from those required for speech.*

A different behavioural hypothesis refers to artistic taste related to the production of objects lacking an immediate functional use or used only for the decoration of the body, which we know happened prior to the Sistine Chapels of Palaeolithic art (Lascaux, Altamira and others). It is very possible that modern humans went from decorating their bodies to doing the same with their homes when caves became permanent living places.

There is no solid evidence detailing how human cognitive capacities gradually improved. The appearance of anatomically modern humans may have represented 'a quantum leap' in cleverness and foresight. This leap may have been related to the advantages of a language characterized by dual patterning: unlimited number of words and unlimited variety of word combinations.

IS CREATIONISM TRUE?

Science and religion need not be in contradiction

'*In the beginning God created the heaven and the earth . . . And God said, Let there be light: and there was light . . . And God called the light Day and the darkness he called Night. And the evening and the morning were the first day . . . So God created man in his own image, in the image of God created he him; male and female created he them . . . And God saw every thing that he had made, and behold, it was very good. And the evening and the morning were the sixth day.*' [7]

The word 'creationism' has many meanings. In its broadest and traditional sense creationism is a religious belief, the idea that a supernatural power, God, created the universe as well as everything that exists in the universe, including humans. In a narrower sense it has come to mean the doctrine that the universe and all that is in it was created by God, essentially in its present form, a few thousand years ago. It is a doctrine that was largely formulated in reaction to Darwin's theory of evolution, based on a literal interpretation of the Bible. Creationism, in this sense, denies the discoveries of astronomy concerning the evolution of the universe and the discoveries of biology concerning the evolution of humans. In the 1990s, particularly in the United States, the notion of 'intelligent design' (ID) was introduced to argue that the design of humans and other organisms cannot be explained by natural processes; but rather they have been designed by an 'intelligent designer' who, implicitly or explicitly, is God, the creator of the universe.

THE CREATION OF ADAM (1508–12) BY MICHELANGELO BUONARROTI.

Are science and religion in opposition?

Scientific knowledge and religious belief need not be in contradiction. If they are correctly assessed, they *cannot* be in contradiction, because they concern non-overlapping realms of knowledge. It is only when assertions are made beyond their legitimate boundaries that science and religious belief appear to be antithetical.

Specifically, does the theory of evolution exclude religious belief? Is it not true that science is fundamentally materialistic and thus excludes any spiritual values? The answer to both questions is 'no'. The scope of science is the world of nature, the reality that is observed, directly or indirectly, by our senses. Science advances explanations concerning the natural world, explanations

> *It is only when assertions are made beyond their legitimate boundaries that science and religious belief appear to be antithetical.*

191

that are subject to the possibility of corroboration or rejection by observation and experiment. Outside that world, it has no authority, no statements to make, no business whatsoever taking one position or another. Science has nothing decisive to say about values, whether economic, aesthetic or moral; nothing to say about the meaning of life or its purpose; nothing to say about religious beliefs (except in the case of beliefs that transcend the proper scope of religion and make assertions that contradict scientific knowledge; such statements cannot be true).

Some scientists, including evolutionists, assert that science denies any valid knowledge concerning values or the world's meaning and purpose. The well-known evolutionary biologist Richard Dawkins explicitly denies design, purpose and values: 'the universe that we observe has precisely the properties we should expect if there is, at bottom, no design, no purpose, no evil and no good, nothing but blind, pitiless indifference'. The historian of science William Provine asserts that 'Modern science directly implies that there are no inherent moral or ethical laws, no absolute guiding principles for human society.'

There is a monumental contradiction implicit in these assertions. If science's commitment to naturalism does not allow it to derive values, meanings or purposes from scientific knowledge, it surely does not allow it to deny their existence either.

In a document published in 1998, *Teaching Evolution and the Nature of Science*, the National Academy of Sciences asserted:

> *Religion and science answer different questions about the world. Whether there is a purpose to the universe or a purpose for human existence are not questions for science . . . Consequently, many people including many scientists, hold strong religious beliefs and simultaneously accept the occurrence of evolution.*

In a similar vein, US Federal Judge John E. Jones III, in a decision dated 20 December 2005 (*Kitzmiller v. Dover Area School District*), wrote:

Many of the leading proponents of ID make a bedrock assumption which is utterly false. Their presupposition is that evolutionary theory is antithetical to a belief in the existence of a supreme being and to religion in general . . . Repeatedly in this trial . . . scientific experts testified that the theory of evolution represents good science, is overwhelmingly accepted by the scientific community, and that it in no way conflicts with, nor does it deny, the existence of a divine creator.

Evolution and the Bible

To some people, the theory of evolution seems incompatible with religious beliefs, particularly those of Christians, because it is inconsistent with the Bible's narrative of creation. The first chapters of the biblical book of Genesis describe God's creation of the world, plants, animals and human beings. A literal interpretation of Genesis seems incompatible with the gradual evolution of humans and other organisms by natural processes.

Many Bible scholars and theologians have long rejected a literal interpretation as untenable, however, because the Bible contains mutually incompatible statements. The very beginning of the book of Genesis presents two different creation narratives. In chapter 1 is the familiar six-day narrative of creation, quoted above, in which God creates human beings, both male and female, after creating light, earth, fish, fowl and cattle. A different narrative appears in chapter 2:

And the Lord God formed man of the dust of the ground . . . And the Lord God planted a garden eastward of Eden; and there he put the man whom he had formed . . . And out of the ground the Lord God formed every beast of the field and every fowl of the air . . . And the Lord God caused a deep sleep to fall upon Adam, and he slept; and he took one of his ribs . . . And the rib, which the Lord God had taken from man, made he a woman, and brought her unto the man.

In this second narrative, Adam is created first, before the Garden of Eden, and before plants and animals. Only afterwards does God create the first woman, out of Adam's rib. Which of the two narratives is correct? Neither one would contradict the other if we would understand them as conveying the same message: that the world was created by God and that humans are His creatures. But both narratives cannot be 'historically and scientifically true' as postulated by the Creation Research Society.

There are numerous inconsistencies and contradictions throughout the Bible – for example in the description of the return from Egypt to the Promised Land by the chosen people of Israel – not to mention erroneous factual statements about the Sun circling around the Earth and the like. Biblical scholars point out that the Bible is infallible with respect to religious truth, not in matters that are of no significance to salvation. Augustine, one of the greatest Christian theologians, wrote in his *De Genesi ad litteram* (*Literal Commentary on Genesis*):

> It is also frequently asked what our belief must be about the form and shape of heaven, according to Sacred Scripture . . . Such subjects are of no profit for those who seek beatitude . . . What concern is it of mine whether heaven is like a sphere and Earth is enclosed by it and suspended in the middle of the universe, or whether heaven is like a disc and the Earth is above it and hovering to one side.

He adds: 'In the matter of the shape of heaven, the sacred writers did not wish to teach men facts that could be of no avail for their salvation.' Augustine is saying that the book of Genesis is not an elementary book of astronomy. Noting that in the Genesis narrative of creation God creates light on the first day but does not create the Sun until the fourth day, he concludes that 'light' and 'days' in Genesis make no literal sense.

Pope John Paul II said in 1981 that the Bible

> speaks to us of the origins of the universe and its makeup, not in order to provide us with a scientific treatise but in order to state the correct relationships of man with God and with the universe. Sacred Scripture wishes simply to declare that the world was

created by God, and in order to teach this truth, it expresses itself in the terms of the cosmology in use at the time of the writer.

Religious fundamentalism in the United States

Opposition to the teaching of evolution in public schools can be traced in the United States to two movements with nineteenth-century roots, Seventh-Day Adventism and Pentecostalism. Consistent with their emphasis on the seventh-day Sabbath as a memorial of the biblical Creation, Seventh-Day Adventists insist on the recent creation of life. This Adventist interpretation of Genesis became the hard core of 'creation science' in the late twentieth century. Many Pentecostalists, who generally endorse a literal interpretation of the Bible, have also adopted and endorsed the tenets of creationism, including the recent origin of Earth. They differ from Seventh-Day Adventists and other creationists in their tolerance of diverse views and the limited import they attribute to the evolution-creation controversy. During the 1920s, biblical fundamentalists helped to persuade more than 20 state legislatures to debate anti-evolution legislation, and four states – Arkansas, Mississippi, Oklahoma and Tennessee – prohibited the teaching of evolution in their public schools.

In 1968, the Supreme Court of the United States declared unconstitutional any law banning the teaching of evolution in public schools (*Epperson v. Arkansas* 393 US97, 1968). Thereafter, Christian fundamentalists introduced legislation in a number of state legislatures ordering that the teaching of 'evolution science' be balanced by allocating equal time to 'creation science'. Creation science, it was asserted, propounds that all kinds of organisms abruptly came into existence when God created the universe, that the world is only a few thousand years old, and that the biblical Flood was an actual event survived by only one pair of each animal species. The legislatures of Arkansas in 1981 and Louisiana in 1982 passed statutes requiring the balanced treatment of evolution science and creation science in their schools, but opponents successfully challenged the statutes as violations of the constitutionally mandated separation of church and state. The

Arkansas statute was declared unconstitutional in federal court in 1982 after a public trial in Little Rock. The Louisiana law was appealed all the way to the Supreme Court of the United States, and in 1987 was ruled unconstitutional on the grounds that, by advancing the religious belief that a supernatural being created humankind, it impermissibly endorsed religion.

More recently, on 28 October 2004, the Dover (Pennsylvania) Area School Board of Directors adopted the following resolution: 'Students will be made aware of gaps/problems in Darwin's theory and of other theories of evolution including, but not limited to, intelligent design.' The constitutional validity of the resolution was challenged in the Federal District Court for the Middle District of Pennsylvania. The trial took place over several weeks, and on 20 December 2005, Federal Judge John E. Jones III issued a 139-page-long decision declaring that 'the Defendants' ID Policy violates the Establishment Clause of the First Amendment of the Constitution of the United States' and that the 'Defendants are permanently enjoined from maintaining the ID Policy'.

Intelligent design

In the 1990s, several authors in the United States advanced a theory of intelligent design (ID), similar to the 'argument from design' advanced over the centuries by Christian authors as a rational demonstration of the existence of God. The argument from design asserts, first, that humans, as well as all sorts of other organisms, have been designed for serving certain functions and for certain ways of life; and second, that only an omnipotent creator could account for the perfection and functional design of living organisms. In 1802, the English clergyman William Paley published the most eloquent and elaborate formulation of the argument. Half a century later, Darwin's theory of natural selection would provide a scientific explanation of the design of organisms.

The ID argument calls for an intelligent designer to explain the supposed irreducible complexity in organisms. An irreducibly complex system is defined by Michael Behe as an entity 'composed of several well-matched, interacting parts that contribute to the basic function, wherein the removal of any one of the parts causes the system to effectively cease functioning'. ID proponents have argued that irreducibly complex systems cannot be the outcome of evolution. According to Behe, 'An irreducibly complex system cannot be produced directly.' Therefore, he affirms, 'If a biological system cannot be produced gradually it would have to arise as an integrated unit, in one fell swoop.'

> *Evolutionists have pointed out again and again, with supporting evidence, that organs and other components of living beings are not 'irreducibly complex' – they do not come about suddenly, or in one fell swoop.*

The argument asserts that unless all parts of the eye, for example, come simultaneously into existence, the eye cannot function; it does not benefit a precursor organism to have just a retina, or a lens, if the other parts are lacking. The human eye, according to this argument, could not have evolved one small step at a time, in the piecemeal manner by which natural selection works.

Evolutionists have pointed out again and again, with supporting evidence, that organs and other components of living beings are not 'irreducibly complex' – they do not come about suddenly, or in one fell swoop. They have shown that the organs and systems claimed by ID proponents to be irreducibly complex are not irreducible at all; rather, less complex versions of the same systems have existed in the past and can be found in today's organisms.

The human eye, the octopus eye or the insect eye did not appear suddenly in their present complexity. Eyes have repeatedly evolved in different animal lineages because sunlight is a pervasive feature of the Earth's environment, to which different animals have adapted in different ways, depending on their physiology and way of life, but always starting from something very simple, even single cells with enzymes sensitive to light (see *What is Natural Selection?*). Scientists have shown that the evolution of eyes has occurred by gradual advancement of the same function – seeing. The process is impelled by natural selection's favouring through time individuals that exhibit functional advantages over others of the same species.

Other instances of alleged irreducible complexity advocated as evidence for ID include the bacterial flagellum, an organ consisting of three components used by some bacteria for swimming; the blood-clotting mechanism in mammals; and the origin of the immune system. How these organs and functions have come about in evolution by natural selection has been satisfactorily explained by scientists. (See, for example, my 2007 book *Darwin's Gift to Science and Religion*, and the references therein.)

Darwin, religion's disguised friend

One difficulty with attributing the design of organisms to a creator is that imperfections and defects pervade the living world. Consider the human eye. The visual nerve fibres in the eye converge to form the optic nerve, which crosses the retina (in order to reach the brain) and thus creates a blind spot, a minor imperfection but an imperfection of design nevertheless; squids and octopuses do not have this defect. Did the designer have greater love for squids than for humans and thus exhibit greater care in designing their eyes than ours?

The theory of ID leads to conclusions about the nature of the designer quite different from those of omniscience, omnipotence and benevolence that Christian theology predicates of God. Organisms and their parts are less than perfect. Deficiencies and dysfunctions are pervasive, evidencing 'incompetent' rather than 'intelligent' design. Consider the human jaw. We have too many teeth for its size, so that wisdom teeth need to be removed and orthodontists make a decent living straightening the others. Would we want to blame God for this blunder? A human engineer would have done better. Evolution gives a good account of this imperfection. Brain size increased over time in our ancestors; the remodelling of the skull to fit the larger brain entailed a reduction of the jaw, so that the head of the newborn would not be too large to pass through the mother's birth canal.

Theologians in the past struggled with the issue of dysfunction because they thought it had to be attributed to God's design. Science, much to the relief of theologians, provides an explanation that convincingly attributes defects, deformities and dysfunctions to natural causes.

More disturbing yet for ID proponents has to be the following consideration. About 20 per cent of all recognized human pregnancies end in spontaneous miscarriage during the first two months of pregnancy. This misfortune amounts at present to more than 20 million spontaneous abortions worldwide every year. Do we want to blame God for the deficiencies in the pregnancy process? Is God the greatest abortionist of them all? Most of us might rather attribute this monumental mishap to the clumsy ways of the evolutionary process than to the incompetence of an intelligent designer.

Examples of deficiencies and dysfunctions in all sorts of organisms can be listed endlessly, reflecting the opportunistic character of natural selection, which achieves imperfect, rather than intelligent, design. The world of organisms also abounds in characteristics that might be called 'oddities', as well as those that have been characterized as 'cruelties', an apposite qualifier if the cruel behaviours were designed outcomes of a being adhering to human or higher standards of morality. However, the cruelties of biological nature are only metaphoric cruelties when applied to the outcomes of natural selection.

Examples of 'cruelty' involve predators tearing apart their prey (say, a small monkey held alive by a chimpanzee biting large morsels from its flesh), or parasites destroying the functional organs of their hosts, but also, and very abundantly, cruelties between organisms of the same species, even between mates. A well-known example is the female praying mantis that devours the male after

> *The theory of evolution, which at first had seemed to remove the need for God in the world, has now convincingly removed the need to explain the world's imperfections as failed outcomes of God's design.*

coitus is completed. A less familiar fact is that, if she gets the opportunity, the female praying mantis will eat the head of the male *before* mating, which thrashes the headless male mantis into spasms of 'sexual frenzy' that allow the female to connect his genitalia with hers. In some midges (tiny flies), the female captures the male as if he were any other prey and injects him with her spittle, which starts digesting the male's innards; the relatively intact male organs, which are partly protected from digestion, break off inside the female and fertilize her. Male cannibalism by their female mates is known in dozens of species, particularly spiders and scorpions. The natural world abounds in 'cruel' behaviours.

Religious scholars in the past struggled with imperfection, dysfunction and cruelty in the living world, things that are difficult to explain if they are the outcome of God's design. The philosopher David Hume expressed the problem succinctly, with brutal directness:

> *Is he [God] willing to prevent evil, but not able? Then he is impotent. Is he able, but not willing? Then he is malevolent. Is he both able and willing? Whence then evil?*

Evolution came to the rescue. Jack Haught, a contemporary Roman Catholic theologian, has written of 'Darwin's gift to theology'. The Protestant theologian Arthur Peacocke has referred to Darwin as the 'disguised friend', by quoting the earlier theologian Aubrey Moore, who in 1891 wrote that 'Darwinism appeared, and, under the guise of a foe, did the work of a friend'. Haught and Peacocke are acknowledging that the theory of evolution, which at first had seemed to remove the need for God in the world, has now convincingly removed the need to explain the world's imperfections as failed outcomes of God's design.

Indeed, a major burden was removed from the shoulders of believers when evidence was advanced that the design of organisms need not be attributed to the immediate agency of the Creator, but rather is an outcome of natural processes. Creationists and proponents of ID would do well to acknowledge Darwin's revolution and accept natural selection as the process that accounts for the design of organisms, as well as for the dysfunctions, oddities and cruelties that pervade the world of life.

Coda: evolution and religion in coexistence

It is possible to believe that God created the world while also accepting that the planets, mountains, plants and animals came about, after the initial creation, by natural processes. As the National Academy of Sciences asserts in the document cited earlier, *Teaching Evolution and the Nature of Science*:

> *Within the Judeo-Christian religions, many people believe that God works through the process of evolution. That is, God has created both a world that is ever-changing and a mechanism through which creatures can adapt to environmental change over time.*

In theological parlance, God may act through secondary causes. Similarly, at the level of the individual, a person can believe he is God's creature without denying that he developed from a single cell in his mother's womb. For the believer, the providence of God impacts personal life and world events through natural causes. The point, once again, is that scientific conclusions and religious beliefs concern different sorts of issues and belong to different realms of knowledge; they do not need to stand in contradiction.

It is possible to believe that God created the world while also accepting that the planets, mountains, plants and animals came about, after the initial creation, by natural processes.

GLOSSARY

allele One of two or more alternative forms of a gene.

amino acids The building blocks of proteins; only 20 are found in proteins although hundreds exist in nature.

anagenesis The evolutionary change of a lineage in the course of time.

analogous Similar in function or appearance, but not due to descent from a common ancestor.

chromosome A threadlike structure, found in the nucleus of cells, that contains the genes arranged in linear sequence; consists of DNA and protein.

chronospecies Organisms of the same lineage that lived at different times and are about as different from one another as are contemporary species.

cladogenesis Split of an evolutionary lineage (or species) into two.

codon A group of three consecutive nucleotides in DNA or mRNA that code for a specific amino acid in a protein.

diploid Having two sets of chromosomes, typically inherited one from each parent.

DNA Deoxyribonucleic acid; the genetic material of most organisms.

enzyme A protein that catalyses a specific chemical reaction.

eukaryote Cell or organism that has the DNA (chromosomes) inside a nucleus.

fitness Reproductive contribution of an individual or genotype to the following generations, usually measured relative to the contribution of other individuals or genotypes.

gamete Sex cell; a mature reproductive cell capable of fusing with a similar cell of the opposite sex; ovules and spermatozoids.

gene Segment of DNA that usually codes for a protein.

gene pool Sum total of the genes of a population or of a species.

genetic code Code that relates each codon to a particular amino acid.

genome Total genetic material of an organism; in diploid organisms it may refer to the half inherited from one parent or to both halves.

genotype Genetic constitution of an individual.

haploid Cell or organism with only one set of chromosomes.

homologous Similarity due to common descent.

heterozygote Having two different alleles.

hybrid Organism resulting from a cross between two genetically different individuals or individuals from different species.

molecular clock Timing of evolutionary events in terms of DNA or protein changes.

molecular evolution Evolution in terms of molecules; usually DNA or proteins.

messenger RNA (mRNA) RNA that conveys DNA information from the cell nucleus to the cytoplasm.

mutation Heritable change in the hereditary molecules, usually genes or proteins.

natural selection Process of differential reproduction of genes or genotypes, and the results of the process.

nucleotide One of four components of DNA or RNA.

nucleus Organelle containing the DNA inside a eukaryotic cell.

phylogeny Sequence of ancestor-descendant forms in an evolutionary lineage.

phylum Major group of organisms: e.g. molluscs, chordates (vertebrates), arthropods (insects).

polyploid Containing more than two sets of chromosomes.

prokaryote A single-cell organism that lacks a membrane-bound nucleus.

protein Molecule made up of amino acids, usually several hundred of 20 different kinds.

RIM Reproductive isolating mechanism.

RNA Ribonucleic acid.

selection coefficient Intensity of selection, usually measured as the reduction in fitness.

species Set of organisms able to interbreed with each other, but not with members of other sets.

taxon Individuals classed together as a group, such as a species, genus, family, order, phylum; plural, taxa.

transcription Transfer of genetic information from DNA to mRNA in the cell nucleus.

translation Transfer of genetic information from mRNA to protein in the cell cytoplasm.

zygote Cell formed by the union (fertilization) of a female gamete (ovule) and a male gamete (spermatozoid).

ENDNOTES

1 *Science, Evolution, and Creationism*, National Academy of Sciences and Institute of Medicine, Washington, D.C. (2008)

2 Charles Darwin, *On the Origin of Species*, 1st edition, ch. 4 (1859)

3 Charles Darwin, *On the Origin of Species*, 5th edition, ch. 3 (1869)

4 *Science, Evolution, and Creationism*, National Academy of Sciences and Institute of Medicine (2008)

5 Charles Darwin, *The Descent of Man* (1871)

6 Charles. Darwin, *The Descent of Man* (1871)

7 *Genesis*, ch. 1, v. 1, 3, 5, 27, 31

INDEX

Quercus Publishing Plc
55 Baker Street
7th Floor, South Block
London,
W1U 8EW

First published in 2012

A catalogue record of this book is available from the British Library

UK and associated territories:
ISBN: 978 1 78087 033 5

Canada:
ISBN: 978 1 84866 192 9

Edited by Jane Selley
Designed and illustrated by Patrick Nugent, except for pages 142 Archives & Special Collections at the Thomas J. Dodd Research Center, University of Connecticut Libraries, and 165 Courtesy Carnegie Institution for Science

Printed and bound in China

10 9 8 7 6 5 4 3 2 1